MOLECULAR NANOMACHINES

ENCYCLOPAEDIA OF NANOSCIENCE

Vol. III

MOLECULAR NANOMACHINES

By

Dr. S.K. Prasad

School of Studies of Zoology & Biotechnology
Vikram University
Ujjain

DPH

DISCOVERY PUBLISHING HOUSE
NEW DELHI-110002

First Published-2007

ISBN 978-81-8356-247-8 (Set)

Published by

DISCOVERY PUBLISHING HOUSE
4831/24, Ansari Road, Prahlad Street,
Darya Ganj, New Delhi-110002 (India)
Phone: 23279245 • Fax: 91-11-23253475
E-mail: dphbooks@rediffmail.com
dphtemp@indiatimes.com

Printed at:

Sachin Printers, Delhi-

PREFACE

There has been rapid progress in nanoscience over the last few years, particularly in the area of miniaturization. The present title **Encyclopaedia of Nanoscience** is the first scientifically detailed description of developments that will revolutionize most of the industrial processes and products currently in use. This ground breaking work draws on physics and chemistry to establish basic concepts and analytical tools and thus only it provides an indispensable introduction to the emerging field of Nanoscience, Scientific research is the ultimate tool in pushing forward the limit of understanding. But, as with any tool, research is only powerful if used properly, and to its full effects. The more we learn, the more we discover connections threading through biochemical and biophysical world. In writing this Encyclopaedia the author has made every effort to present these connections in a way that will help first time students of Nanoscience understand the subject and how very relevant it is to their lives.

In the preparation of this book large number of books and research papers have been consulted. So no authenticity is claimed.

The author expresses his gratitude to Mr. Wasan and staff of M/s Discovery Publishing House for their whole hearted co-operation in the publication of this book.

The author tried hard to be accurate and upto date in statement and realises the impossibility of completely avoiding errors therefore, the author will greatly appreciate having his attention called to any questionable statement.

Author

Contents

1

INTRODUCTION

It is crucial to realize that the physics on the *nanometer scale* tends to become dominated by quantum physics. In the nanoworld one must always be prepared to take seemingly strange quantum phenomena into account and hence give up on an entirely classical description. Although this is not a course in quantum physics it is nevertheless imperative to get a grasp of the basic ideas and concepts of quantum physics. Without this it is not possible to reach a full understanding of the potentials of nanotechnology. Serious students of nanotechnology are hereby encouraged to study at least a minimum of quantum theory.

PARTICLE-WAVE DUALITY

The first laws of quantum physics dealt with energy quantization. They were discovered in studies of the electromagnetic radiation field by Planck and Einstein in 1900 and 1905, respectively. A new universal constant, *Planck's constant h*, was introduced in physics in addition to other constants like the speed of light, c, the gravitation constant, G, and the charge quantum, e. The 1998 CODATA values for h and $\hbar = h/2\pi$ are

$$h = 6.62606876(52) \times 10^{-34} \text{ Js}, \qquad \text{...(1a)}$$

$$\hbar = 1.054571596(82) \times 10^{-34} \text{ Js}. \qquad \text{...(1b)}$$

The energy E of light quanta (*photons*) in light of frequency f or angular frequency $\omega = 2\pi f$ is given by

$$E = hf = \hbar\omega. \qquad \text{...(2)}$$

Already in 1906 Einstein applied this energy quantization on other objects than light, namely on oscillating atoms in his work on the heat capacity of solids. But it was first with *Bohr's analysis* of the stationary states in the hydrogen atom that quantum physics really proved to be essential for the understanding of not only the radiation field, but also of matter. With Heisenberg and Schrodinger's seminal papers from 1925 and 1926, respectively, modern quantum theory was born, a theory that eversince has been the foundation of our understanding of the physical world in which we live.

A central concept in quantum physics is the particle-wave duality, the fact that fundamental objects in the physical world, *electrons*, *protons*, *neutrons*, *photons* and other *leptons*, *hadrons*, and *field quanta*, all have the same dual nature: they are at the same time both particles and waves. In some situations the particle aspect may be the dominant feature, in other *vice versa*; but the behaviour of any given object can never be understood fully by ascribing only one of these aspects to it.

Historically, as indicated, the particle-wave duality was first realized for the electromagnetic field. It is interesting to note that right from the beginning when the first theories of the nature of light was proposed in the 17th century, it was debated whether light were *particles* (*corpuscles*), as claimed by Newton, or *waves*, as claimed by Huygens. The debate appeared to end in the beginning of the 19th century with Young's famous double-slit interference experiments that demonstrated that light were *waves*. With Maxwell's theory (1873) and Hertz's experiments (1888) the light waves were shown to be of electromagnetic nature. But after nearly one hundred years of wave dominance Planck's formula for the energy distribution in black-body radiation demonstrated that light possesses some element of particle nature. This *particle aspect* became more evident with Einstein's explanation of the photoelectric effect: small energy parcels of light, the so-called *light quanta*, are able to knock out electrons from metals just like one billiard ball hitting another. For some years theorists tried to give alternative explanations of the photoelectric effect using maxwellian waves, but it proved impossible to account for the concentration of energy in a small point needed to explain the photoelectric effect without postulating the existence of light quanta—today called *photons*.

In 1913 Niels Bohr published his theory of the *hydrogen atom*, explaining its stability in terms of stationary states. Bohr did not explain why stationary states exist. He boldly postulated their existence and from that assumption he could explain the frequencies of the experimentally observed spectral lines, and in particular he could derive Balmer's empirical expression for the position of the spectral lines.

He also derived a formula for the *Rydberg constant* appearing in that expression. In the following years it was postulated still without an explanation that a particle of momentum $p = mv$, where m and v is its mass and velocity, respectively, moving in a closed orbit must obey the *Bohr-Sommerfeld quantization rule*, $\oint p \cdot dr = nh$, where the integral is over one revolution and n is an integer.

In 1923 de Broglie proposed that particles could be ascribed a *pilot wave* guiding their motion through space. This idea was based on the fundamental idea of duality: if light waves were also particles, then ought not also particles be waves? Moreover, if particles were waves it would somehow be possible to explain Bohr's stationary states as a kind of standing particle-waves. Think of a vibrating string fixed in both ends: only certain resonance frequencies are possible corresponding to matching an integral number of half-waves between the endpoints. Likewise for higher dimensional bodies.

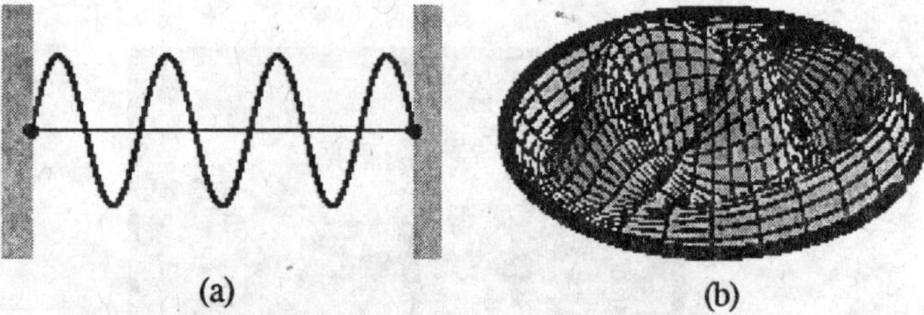

Fig. 1.1. Resonance modes or eigenmodes in one and two dimensions: (a) a vibrating string, and (b) a vibrating membrane.

Below we shall go through a simple argument that leads to de Broglie's famous relation between the momentum p of a particle and the wave length λ (or wave number $k = 2\pi/\lambda$) of its associated wave,

$$p = \frac{h}{\lambda} = \hbar k. \qquad ...(3)$$

Shortly after de Broglie's proposal Davisson and Germer verified his ideas by demonstrating the existence of diffraction patterns when electrons are shot through a thin metal film. The regularly spaced atoms in the film constituted a multi-slit analogue of Young's double-slit experiment for light. In the beginning of 1926 Schrodinger published his wave equation providing a firm mathematical foundation for de Broglie's ideas.

Today there can be no doubt about the reality of the wave nature of particles. Many experiments show this convincingly. In fact, all our understanding of matter on the microscopic level is based on the

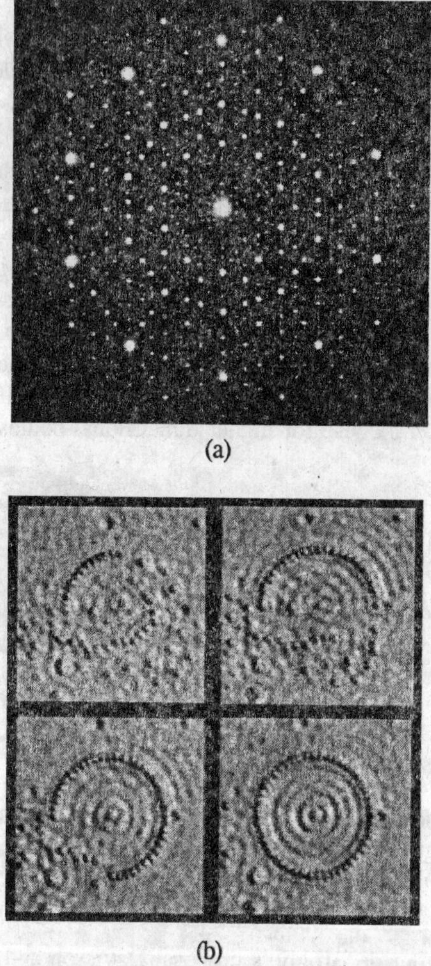

(a)

(b)

Fig. 1.2. Two strong evidences for the existence of electron waves. (a) An electron diffraction from the quasicrystal $Al_{70}Co_{11}Ni_{19}$; (b) Electron waves on the surface of copper detected by scanning tunnel microscopy.

existence of these waves. Two particularly beautiful observations of electron waves are shown in Fig 1.2.

de Broglie Waves

Consider a given particle of mass m moving along the x axis with momentum p. If a wave $\psi(x, t)$ is to be associated with this motion, what is then the relation between the wavelength λ and p? In the

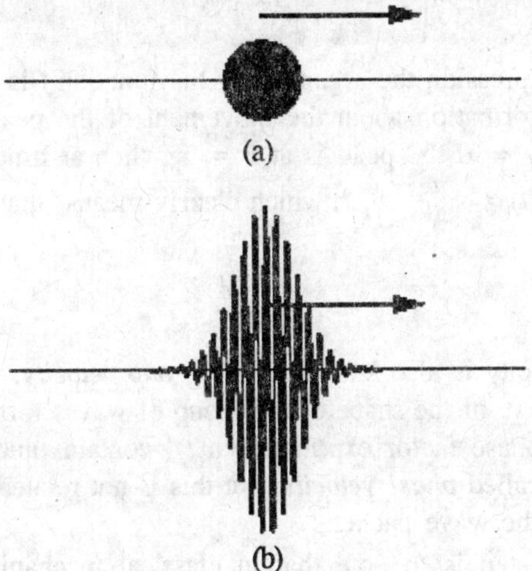

Fig. 1.3. (a) A particle moving with speed v along the x-axis. (b) A wave packet with a center motion identical to that of the particle.

following we shall show that the answer is the de Broglie relation Eq. (3), stating that $p = h/\lambda = \hbar k$.

For the wave description to be reasonable we demand that the wave has a high amplitude at the position x of the particle, i.e. $\psi(x, t)$ must be a wave packet that moves with the velocity $v = p/m$. Such a wave packet can be written in the form

$$\psi(x,t) = \int_{-\infty}^{\infty} dk g(k) e^{i(kx-\omega_k t)} \qquad \text{...(4)}$$

where the weight function $g(k)$ is peaked around $k = k_0$. In these notes $i = \sqrt{-1}$. The wider $g(k)$ is in k space the narrower $\psi(x, t)$ is in x space. An example of a wave packet with $g(k) = \exp[-L^2(k - k_0)^2]$ is shown in Fig. 1.3. The k-dependence of the frequency ωk is called the *dispersion* relation for the particle. If $g(k)$ is sufficiently narrow, we can expand the dispersion relation around $k = k_0$,

$$\omega k \approx \omega_0 + \frac{\partial \omega}{\partial k}(k - k_0) \qquad \text{...(5)}$$

where $\omega_0 = \omega k_0$. We insert this expansion in Eq. (4) and use $q = k - k_0$ to obtain

$$\psi(x,t) \approx \int_{-\infty}^{\infty} dk\ g(k) e^{i\left[kx - \left(\omega_0 + \frac{\partial \omega}{\partial k}(k-k_0)\right)t\right]} \qquad \text{...(6a)}$$

$$= \left[\int_{-\infty}^{\infty} dq\ g(k_0 + q) e^{iq\left[x - \frac{\partial \omega}{\partial k}t\right]}\right] e^{i(k_0 x - \omega_0 t)} \qquad \text{...(6b)}$$

$$= f\left(x - \frac{\partial \omega}{\partial k}t\right)e^{i(k_0 x - \omega_0 t)} \qquad \text{...(6c)}$$

In this expression the argument of the function f is the only place containing information about the movement of the peak of the *wave packet*. If at $t = 0$ the peak is at $x = x_0$, then at time t its position is given by $x(t) = x_0 + \frac{\partial \omega}{\partial k}t$, which clearly means, that its velocity υ is given by

$$\upsilon = \frac{\partial \omega}{\partial k} \qquad \text{...(7)}$$

This velocity is also known as the *group velocity*, since it refers to the movement of the shape of the group of waves forming the wave packet. The phase factor $\exp[i(k_0 x - \omega_0 t)]$ contains another velocity, ω_0/k_0, the so-called *phase velocity*, but this is not related in anyway to the shape of the wave packet.

The last step is to note that in classical mechanics the *kinetic energy* is given by

$$E = \frac{1}{2}m\upsilon^2 = \frac{p^2}{2m}.$$

From this and from Eqs (2) and (7) we find

$$\upsilon = \frac{p}{m} = \frac{\partial E}{\partial p} \qquad \text{...(8a)}$$

$$\upsilon = \frac{\hbar \partial \omega}{\hbar \partial k} = \frac{\partial E}{\partial (\hbar k)} \qquad \text{...(8b)}$$

These equations become consistent if we make the following identification,

$$p = \hbar k = \frac{h}{\lambda} \qquad \text{...(9)}$$

which in fact is the de Broglie relation.

The wave picture leads to a very fundamental relation between the uncertainty in position Δx and the uncertainty in wave number Δk. Here the uncertainty Δk in k around k_0 is of the order $1/L$ as seen from the gaussian $g(k) = \exp[-L^2(k - k_0)^2]$, while the uncertainty in x is of the order $2L$ as seen from the gaussian $\exp(-x^2/4L^2)$. The more well defined the position is the less well defined is the wave vector and vice versa. This is summarized in the inequality $\Delta x \ \Delta k \geq 1$, which after multiplication with $\hbar$ becomes the famous Heisenberg uncertainty relation

$$\Delta x\, \Delta p \geq \hbar \qquad ...(10)$$

In the extreme quantum limit, e.g. for the ground state of a system the inequality becomes an equality. Furthermore, Δx becomes the characteristic length x_c, while Δp becomes the characteristic momentum p_c:

$$p_c = \frac{\hbar}{x_c} \qquad ...(11)$$

Quantum Pressure

Let us immediately focus on a very essential physical consequence of the wave nature of matter contained in the *de Broglie relation*, namely the appearance of a *quantum pressure*. Imagine we want to localize a particle by putting it in a littlc cubic box wilh side length L. Mathematically, localization is described by demanding $\psi = 0$ at the walls of the cube and everwhere outside it. The longest possible wave length λ_{max} of the particle in the box is given by the half wave length condition $\lambda_{max}/2 = L$. The smallest possible kinetic energy of the particle in the box is thus given by

$$E_{min} = \frac{p_x^2 + p_y^2 + p_z^2}{2m} = \frac{3}{2m}\left(\frac{h}{\lambda_{max}}\right)^2 = \frac{3h^2}{8mL^2} \xrightarrow[L\to 0]{} \infty \qquad ...(12)$$

We see that it costs energy to localize a particle in a little region of space. The somewhat strange quantum nature of a particle reveals itself as a very concrete pressure acting against localization of the particle. This is the so-called *quantum pressure*, a very central concept in modern physics. Among other things it determines the size of atoms, the speed of sound in solids, the height of mountains on the Earth, and the size of neutron stars. Since $E_{min} \propto L^{-2} = V^{-\frac{2}{3}}$ the quantum pressure Pqu becomes

$$P_{qu} = \frac{\partial E_{min}}{\partial V} = \frac{2}{3}\frac{E_{min}}{V} \qquad ...(13)$$

As studied already the combination of quantum pressure and electrostatics defines the size of *atoms*. If the characteristic radius of the hydrogen atom is denoted a we can use of the Heisenberg equality of Eq. (11) with $x_c = a$ to rewrite the expression for the classical energy as follows:

$$E(a) = \frac{p^2}{2m} - \frac{e^2}{4\pi\varepsilon_0}\frac{1}{a} \simeq \frac{\hbar^2}{2m}\frac{1}{a^2} - \frac{e^2}{4\pi\varepsilon_0}\frac{1}{a} \qquad ...(14)$$

Finding the size a_0 that minimizes $E(a)$ we arrive (a bit by chance given the approximation) at the famous expression for the Bohr radius of the hydrogen atom,

$$a_0 = \frac{4\pi\varepsilon_0\hbar^2}{me^2} = 0.053\ nm \quad ...(15)$$

and for the ground state energy $E(a_0)$,

$$E_0 = E(a_0) = \frac{1}{2}\frac{1}{4\pi\varepsilon_0}\frac{e^2}{2ma_0^2} = -\frac{me^4}{2(4\pi\varepsilon_0)^2\hbar^2} = -13.6\ eV \quad ...(16)$$

SCHRODINGER EQUATION IN ONE-DIMENSION

In 1926 Schrodinger wrote down the first proper wave equation for *de Broglie's matter waves*. The time-dependent Schrodinger equation for the wave function $\psi(x, t)$ describing a single non-relativistic particle of mass m moving in one dimension in a potential $V(x)$ is

$$i\hbar\frac{\partial}{\partial t}\psi(x,t) = -\frac{\hbar^2}{2m}\frac{\partial^2}{\partial x^2}\psi(x,t) + V(x)\psi(x,t) \quad ...(17)$$

The Schrodinger equation forms the foundation of quantum physics, and as such it cannot logically be derived from classical physics. However, we will nevertheless give a heuristic derivation of it in the following.

A *wave equation* for a given field is a partial differential equation that describes the local evolution of the field in any given space-time point (x, t) in terms of the partial derivatives $\frac{\partial}{\partial t}$ and $\frac{\partial}{\partial x}$. Since quantum behaviour through Eqs. (2) and (3) has been linked to energy, $E = \hbar\omega$ and momentum $p = \hbar k$, it is natural to seek a wave equation corresponding to the classical energy expression containing both these quantities,

$$E = \frac{p^2}{2m} + V(x). \quad ...(18)$$

But how do we get the partial derivatives into play? And how do we take the spatially varying potential into account? Well, the clue comes from considering at first a pure plane wave, $\psi_0(x, t) = e^{i(kx-\omega t)}$, together with the quantum conditions Eqs. (2) and (3):

$$i\hbar\frac{\partial}{\partial t}e^{i(kx-\omega t)} = \hbar\omega\ e^{i(kx-\omega t)} = E\ e^{i(kx-\omega t)} \quad ...(19a)$$

$$\frac{\hbar}{i}\frac{\partial}{\partial x}e^{i(kx-\omega t)} = \hbar k\ e^{i(kx-\omega t)} = p\ e^{i(kx-\omega t)}. \quad ...(19b)$$

This result is then generalized to the case of any shape of the wave function $\psi(x, t)$ by making the fundamental assumption of quantum theory, namely that the *classical energy E* and *momentum p* are replaced with the following differential operators, the so-called *Hamiltonian operator* $\hat{H}$ and the *momentum operator* $\hat{p}$:

$$E \rightarrow \hat{H} = i\hbar\frac{\partial}{\partial t}, \quad \text{...(20a)}$$

$$p \rightarrow \hat{p} = \frac{\hbar}{i}\frac{\partial}{\partial x}. \quad \text{...(20b)}$$

The Schrodinger equation Eq. (17) now follows by direct substitution,

$$E = \frac{p^2}{2m} + V(x) \rightarrow \hat{H} = \frac{\hat{p}}{2m} + V(x) \rightarrow i\hbar\frac{\partial}{\partial t} = \frac{\hbar^2}{2m}\frac{\partial^2}{\partial x^2} + V(x). \quad \text{...(21)}$$

Schrodinger Equation in Three-dimensions

It is a simple matter to extend the results of the previous section to three spatial dimensions and write down the corresponding time-dependent *Scrodinger equation*. Using cartesian coordinates, positions are given by $r = (x, y, z)$ and momenta by $p = (p_x, p_y, p_z)$. We derive the Schrodinger equation for a particle of mass m moving in the potential $V(x, y, z)$ by applying the quantum rule Eq. (20b) for each spatial direction. The energy

$$E = \frac{p_x^2 + p_y^2 + p_z^2}{2m} + V(x,y,z) \quad \text{...(22)}$$

becomes the time-dependent Schrodinger equation,

$$i\hbar\frac{\partial}{\partial t}\psi(x,y,z,t) = \left[-\frac{\hbar}{2m}\left(\frac{\partial^2}{\partial x^2} + \frac{\partial^2}{\partial y^2} + \frac{\partial^2}{\partial z^2}\right) + V(x,y,z)\right]\psi(x,y,z,t). \quad \text{...(23)}$$

This is often written in the abbreviated form,

$$i\hbar\frac{\partial}{\partial t}\psi(r,t) = \left[-\frac{\hbar^2}{2m}\nabla^2 + V(r)\right]\psi(r,t), \quad \text{...(24)}$$

and the momentum operator $\hat{p}$ itself as

$$\hat{p} = \frac{\hbar}{i}\left(\frac{\partial}{\partial x}, \frac{\partial}{\partial y}, \frac{\partial}{\partial z}\right) = \frac{\hbar}{i}\nabla. \quad \text{...(25)}$$

Often the symmetry of a given problem makes it obvious which particular coordinate system to choose. For example, the *Coulomb potential* around a single proton is spherical symmetric hence making *spherical coordinates* (r, θ, ϕ) the natural choise in studies of the electron wave functions in the hydrogen atom. The form of the differential operators nabla, ∇, and the Laplacian, ∇^2, in Cartesian, cylindrical, and spherical coordinates are the following:

(a) Cartesian coordinates with basis vectors e_x, e_y, and e_z:

$$\nabla = e_x \frac{\partial}{\partial x} + e_y \frac{\partial}{\partial y} + e_z \frac{\partial}{\partial z}, \qquad \text{...(26a)}$$

$$\nabla^2 = \frac{\partial^2}{\partial x^2} + \frac{\partial^2}{\partial y^2} + \frac{\partial^2}{\partial z^2}. \qquad \text{...(26b)}$$

(b) Cylindrical coordinates with basis vectors e_r, e_ϕ, and e_z:

$$\nabla = e_r \frac{\partial}{\partial r} + e_\phi \frac{1}{r}\frac{\partial}{\partial \phi} + e_z \frac{\partial}{\partial z}, \qquad \text{...(27a)}$$

$$\nabla^2 = \frac{1}{r}\frac{\partial}{\partial r}\left(r\frac{\partial}{\partial r}\right) + \frac{1}{r^2}\frac{\partial^2}{\partial \phi^2} + \frac{\partial^2}{\partial z^2}. \qquad \text{...(27b)}$$

(c) Spherical coordinates with basis vectors e_r, e_θ, and e_ϕ:

$$\nabla = e_r \frac{\partial}{\partial r} + e_\theta \frac{1}{4}\frac{\partial}{\partial \theta} + e_\phi \frac{1}{r\sin\theta}\frac{\partial}{\partial \phi}, \qquad \text{...(28a)}$$

$$\nabla^2 = \frac{1}{r^2}\frac{\partial}{\partial r}\left(r^2\frac{\partial}{\partial r}\right) + \frac{1}{r^2 \sin\theta}\frac{\partial}{\partial \theta}\left(\sin\theta\frac{\partial}{\partial \theta}\right) + \frac{1}{r^2\sin^2\theta}\frac{\partial^2}{\partial \phi^2} \qquad \text{...(28b)}$$

SUPERPOSITION AND INTERFERENCE OF QUANTUM WAVES

The Schrodinger equation is a linear differential equation. This means that if ψ_1 and ψ_2 are two solutions of it, then so is the sum $\psi = \psi_1 + \psi_2$. In other words, quantum waves obey the superposition principles in analogy with electromagnetic waves. It follows immediately that interference is possible,

$$|\psi|^2 = |\psi_1 + \psi_2|^2 = |\psi_1|^2 + |\psi_2|^2 + 2\,\mathrm{Re}(\psi_1^* \psi_2). \qquad \text{...(29)}$$

We see that the total probability density $|\psi|^2$ is not merely the sum of the two partial probability densities $|\psi_1|^2$ and $|\psi_2|^2$. The interference term $2\mathrm{Re}(\psi_1^* \psi_2)$ has to be taken into account, and the measurement of interference effects like the electron diffraction pattern is crucial in demonstrating the existence of quantum waves.

Another consequence of the linearity of the *Schrodinger equation* is the abstract formulation of quantum physics in terms of a particular vector space, the so-called *Hilbert space*. Each wave function can be thought of as a vector in this abstract vector space. Addition of these vectors are possible due to the superposition principle. We refer the reader to any standard text on quantum mechanics for further studies of the Hilbert space formulation of quantum theory.

Energy Eigenstates

A particularly important class of wave functions are the so-called *energy eigenfunctions*. They are characterized by having a well defined energy at all times, i.e. they correspond to Bohr's idea of stationary states. We try to find solutions of the form

$$\psi(r,t) = \psi_E(r)\, e^{-i\frac{E}{\hbar}t}. \qquad ...(30)$$

Almost by definition this wave function can be ascribed the energy E since

$$i\hbar\frac{\partial}{\partial t}\psi_E(r)\, e^{-i\frac{E}{\hbar}t} = E\psi_E(r)\, e^{-i\frac{E}{\hbar}t}. \qquad ...(31)$$

If we insert this particular form of $\psi(r, t)$ into the time-dependent Schrodinger equation Eq. (17) and divide out the common factor $\exp[-i\frac{E}{\hbar}t]$, we end up with the so-called *time-independent* Schrodinger equation,

$$\left[-\frac{\hbar^2}{2m}\nabla^2 + V(r)\right]\psi_E(r) = E\psi_E(r). \qquad ...(32)$$

Eigenvalue problems like the time-independent Schrodinger equation Eq. (31) are solved by finding both the wavefunction $\psi_E(r)$ and the corresponding eigenenergy E. Often, the set of possible values E is restricted to a discrete set. The state $\psi_E(r)$ is seen to be the stationary states originally proposed by Bohr.

Interpretation of the Wavefunction Ψ

Let us now address the question of how to interpret the wavefunction that now has been associated with a given particle. It would certainly be desirable if the wavefunction ψ has a large amplitude where the particle is present, and a vanishing amplitude where the particle is absent. We can be guided from our experience with light and photons.

Intensity Argument

The *wavefunction* for light is the electric field $E(r, t)$, while the intensity $I(r, t)$ is proportional to the square of the electric field. On the other hand, using the photon representation of light, we find that the intensity is proportional with the local number of photons per volume, the photon density $n(r, t)$,

$$n(r,t) \propto I(r,t) \propto |E(r,t)|^2. \qquad ...(33)$$

Naturally, the total number of photons $N(t)$ at time t is found by integrating over the entire volume V,

$$N(t) = \int dr\ n(r,t). \qquad ...(34)$$

Returning to the quantum wavefunction $\psi(r, t)$ of a single particle we postulate in analogy with Eq. (33) that the intensity $|\psi(r, t)|^2$ of the wavefunction is related to the particle density n and write

$$n(r,t) \propto I(r,t) \propto |\psi(r,t)|^2. \qquad ...(35)$$

But since we have only one particle the analogy of Eq. (34) reduces to

$$1 = \int dr\ n(r,t). \qquad ...(36)$$

This looks like the condition for a probability distribution. The total probability for finding the particle somewhere in space at time t is indeed unity. In fact, the particle and wave description can be reconciled by the fundamental postulate of quantum physics:

$|\psi(r, t)|2\ dr$ is the probability that the particle is in the volume dr around the point r at time t.

$\psi(r, t)$ itself is denoted the probability amplitude.

The probability interpretation of the wave function means that the wave function ψ must always be normalized to unity. Therefore all wavefunctions $C\psi$ proportional to ψ represent the same physical state.

Continuity Equation Argument

The probability interpretation can be supported by the following analysis. If $n(r, t) = |\psi(r, t)|^2$ really is a *probablity density*, then, since a non-relativistic particle as the one studied here cannot disappear, it must obey a continuity equation

$$\frac{\partial n}{\partial t} = -\nabla \cdot J. \qquad ...(38)$$

Here J is some probability current density to be determined. Such an equation is well-known in *electromagnetism*, where n is the electric charge density and J is the electric current density. The probability current density J is found using $n = |\psi|^2 = \psi^*\psi$ and the Schrodinger equation as follows. Begin by calculating the time derivative:

$$\frac{\partial n}{\partial t} = \frac{\partial \psi^* \psi}{\partial t} = \psi^* \frac{\partial \psi}{\partial t} + \frac{\partial \psi^*}{\partial t}. \qquad ...(39)$$

Then write down the Schrodinger equation for ψ and its complex conjugate ψ^*,

$$\frac{\partial \psi}{\partial t} = +\frac{1}{i\hbar}\left[-\frac{\hbar^2}{2m}\nabla^2\psi + V\psi\right], \quad \text{...(40a)}$$

$$\frac{\partial \psi^*}{\partial t} = \frac{1}{i\hbar}\left[-\frac{\hbar^2}{2m}\nabla^2\psi^* + V\psi^*\right]. \quad \text{...(40b)}$$

The equation for ψ is multiplied by ψ^*, and the equation for ψ^* is multiplied by ψ. The resulting equations are added. The V-terms cancel and using Eq. (39), leads to

$$\frac{\partial \psi^*\psi}{\partial t} = -\frac{\hbar}{2mi}\left[\psi^*\nabla^2\psi - \psi\nabla^2\psi^*\right] \quad \text{...(41)}$$

Now it is a simple matter to rewrite this as a divergence,

$$\frac{\partial \psi^*\psi}{\partial t} = \frac{\hbar}{2mi}\nabla\cdot\left[\psi^*\nabla\psi - \psi\nabla\psi^*\right] \quad \text{...(42)}$$

Furthermore, it is tempting to bring in the momentum operator $\hat{p} = (\hbar/i)\nabla$, which yields

$$\frac{\partial \psi^*\psi}{\partial t} = -\nabla\cdot\left[\frac{1}{2m}\left\{\psi^*\left(\frac{\hbar}{i}\nabla\psi\right)^*\right\}\right] = -\nabla\cdot\left[\frac{1}{2m}\left\{\psi^*(\hat{p}) + \psi(\hat{p}\psi)^*\right\}\right]. \quad \text{...(43)}$$

We finally arrive at

$$\frac{\partial \psi^*\psi}{\partial t} = -\nabla\cdot\frac{1}{m}\mathrm{Re}\left[\psi^*(\hat{p}\psi)\right] \quad \text{...(44)}$$

The continuity equation Eq. (38) for the probability density is established if we simply identify J as

$$J(r,t) = \frac{1}{m}\mathrm{Re}\left[\psi^*(\hat{p}\psi)\right] \quad \text{...(45)}$$

Fortunately for the probability interpretation this is a very reasonable result. Recall that the classical expression for current density is $J = n\,v = n\,p/m$. Given that $n = \psi^*\psi$, a naive guess for the quantum expression would be $J = \psi^*\psi\,p/m$, which in fact is not far from the exact result Eq. (45). The latter is just slightly more complicated due to $\hat{p}$ being a differential operator and not a number.

Quantum Operators and their Expectation Values

We have already met some quantum operators, e.g. the *energy operator* $\hat{H} = i\hbar\frac{\partial}{\partial t}$, the *momentum operator* $\hat{p} = \frac{\hbar}{i}\nabla$, and the *potential*

operator $\hat{V} = V(r)$. When an operator O acts on a given wavefunction ψ the result is in general not proportional to the wavefunction, $O\psi \neq$ = const ψ. How should one find the value of the operator? The answer comes from probability theory. Since the wavefunction has been identified with a probability amplitude, one is reminded of how mean values are calculated in classical probability theory. If X is some stochastic variable and $P(X)$ is the probability distribution of X, then the mean value or expectation value $f(X)$ of any function f of the stochastic variable is given by

$$(f(X)) = \int dX\ f(x)\ P(X). \qquad ...(46)$$

Since ψ loosely speaking is the squareroot of P we arrive at the following definition of the expectation value $(\hat{O})_\psi$ of quantum operator O acting on the state ψ:

$$(\hat{O})_\psi = \int dr\ \psi^*(r)\hat{O}\psi(r). \qquad ...(47)$$

We shall use this equation many times.

MANY-PARTICLE QUANTUM STATES

In general a physical system contains more than one particle, and in that case we need to extend the wavefunction formalism.

N-particle Wavefunction

Consider a system containing N identical particles, say, *electrons*. Let the positions of the particles be given by the N vectors r_1, r_2, ...r_N. The wavefunction of the system is given by $\psi(r_1, r_2, ..., r_N)$, which is a complex function in the $3N$-dimensional configuration space. The first fundamental postulate for many-particle systems is to interpret the N-particle wavefunction as a probability amplitude such that its absolute square is related to a probability,

$|\psi((r_1, r_2, ..., r_N)|^2 \prod_{j=1}^{N} dr_j$ is the probability that the N particles are in the $3N$-dimensional volume $\prod_{j=1}^{N} dr_j$ surrounding the point $(r_1, r_2, ..., r_N)$ in the $3N$-dimensional configuration space. ...(48)

Permutation Symmetry and Indistinguishability

A fundamental difference between classical and quantum mechanics concerns the concept of indistinguishability of identical particles. In *classical mechanics* each particle in an many-particle system can be equipped with an identifying marker (e.g. a coloured spot on a billiard ball) without influencing its behaviour, and moreover it follows its own continuous path in phase space. Thus in principle each particle in a group of identical particles can be identified. This is not so in quantum mechanics. Not even in principle is it possible to mark a particle without influencing its physical state; and worse, if a number of identical particles are brought to the same region in space, their wavefunctions will rapidly spread out and overlap with one another, thereby soon render it impossible to say which particle is where. It has for example no strict physical meaning to say that particle 1 is in state $\psi_{\nu 1}$ and particle 2 is in state $\psi_{\nu 2}$, instead we can only say that two particles are occupying the two states $\psi_{\nu 1}$ and $\psi_{\nu 2}$. The second fundamental assumption for N-particle systems is therefore that

Identical particles, i.e. particles characterized by the same quantum numbers such as mass, charge and spin, are in principle indistinguishable. ...(49)

From the indistinguishability of particles follows that if two coordinates in an N-particle state function are interchanged the same physical state results, and the corresponding state function can at most differ from the original one by a simple prefactor C. If the same two coordinates then are interchanged a second time, we end with the exact same state function,

$$\psi(r_1,..,r_j,..,r_k,..,r_N) = C\psi(r_1,..,r_k,..,r_j,..,r_N)$$
$$= C^2\psi(r_1,..,r_j,..,r_k,..,r_N), \quad ...(50)$$

and we conclude that $C^2 = 1$ or $C = \pm 1$. Only two species of particles are thus possible in quantum physics, the so-called *bosons* and *fermions*:

$$\psi(r_1,\ldots,r_j,\ldots,r_k,\ldots,r_N) = +\psi(r_1,\ldots,r_k,\ldots,r_j,\ldots,r_N) \text{ (bosons)}, \quad ...(51a)$$

$$\psi(r_1,\ldots,r_j,\ldots,r_k,\ldots,r_N) = -\psi(r_1,\ldots,r_k,\ldots,r_j,\ldots,r_N) \text{ (fermions)} \quad ...(51b)$$

The importance of the assumption of indistinguishability of particles in quantum physics cannot be exaggerated, and it has been introduced due to overwhelming experimental evidence. For fermions it immediately leads to the Pauli exclusion principle stating that two fermions cannot occupy the same state, because if in Eq. (51b) we let $r_j = r_k$ then $\psi = 0$ follows. It thus explains the periodic table of the elements, and consequently forms the starting point in our understanding of atomic physics, condensed matter physics and chemistry. It furthermore plays a fundamental role in the studies of the nature of stars and of the scattering processes in high energy physics. For bosons the assumption is necessary to understand Planck's radiation law for the electromagnetic field, and spectacular phenomena like Bose–Einstein condensation, superfluidity and laser light.

Fermions: Wavefunctions and Occupation Number

Consider two identical fermions, e.g. *electrons*, that are occupying two single-particle states $\psi_{\nu 1}$ and $\psi_{\nu 2}$. It is useful to realize how to write down the two-particle wavefunction $\psi_{\nu 1 \nu 2}(r_1, r_2)$ that explicitly fulfill the fermionic antisymmetry condition Eq. (51b). It is easy to verify that the following function will do the job,

$$\psi_{\nu_1\nu_2}(r_1,r_2) = \frac{1}{\sqrt{2}}\left[\psi_{\nu_1}(r_1)\psi_{\nu_2}(r_2) - \psi_{\nu_1}(r_2)\psi_{\nu_2}(r_1)\right] \quad ...(52)$$

where the prefactor is inserted to ensure the correct normalization:

$$\int dr_1 \int dr_2 \left|\psi_{\nu_1\nu_2}(r_1,r_2)\right|^2 = 1. \quad ...(53)$$

In the case of N fermions occupying N orbitals $\psi_{\nu 1}, \psi_{\nu 2} \ldots \psi_{\nu N}$ the operator $\hat{S}_-$ that antisymmetrizes the simple product of the N single-particle wavefunctions is the so-called Slater determinant

$$\psi_{\nu_1 \ldots \nu_N}(r_1,r_2,\ldots,r_N) = \hat{S}_-\left[\psi_{\nu_1}(r_1)\psi_{\nu_2}(r_2)\ldots\psi_{\nu_N}(r_N)\right] \quad ...(54a)$$

$$= \frac{1}{\sqrt{N!}}\begin{vmatrix} \psi_{\nu_1}(r_1) & \psi_{\nu_1}(r_2) & \cdots & \psi_{\nu_1}(r_N) \\ \psi_{\nu_2}(r_1) & \psi_{\nu_2}(r_2) & \cdots & \psi_{\nu_2}(r_N) \\ \vdots & \vdots & \ddots & \vdots \\ \psi_{\nu_N}(r_1) & \psi_{\nu_N}(r_2) & \cdots & \psi_{\nu_N}(r_N) \end{vmatrix}. \quad ...(54b)$$

Now that we know a little about what the fermion N-particle states look like, we turn to the question of what is the probability for having a given orbital occupied at *non-zero temperature*. Let n_ν be the occupation number for the state ψ_ν. At zero temperature n_ν is either constantly 0 or constantly 1. As the temperature T is raised from zero n_ν begins to fluctuate between 0 and 1. The thermal mean value (n_ν) is denoted $f_F(\varepsilon_\nu)$, where ε_ν is the eigenenergy of ψ_ν. From statistical mechanics we know that the for a system with a fluctuating particle number thermal averages are calculated using the weight factor exp[-$\beta(n_\nu\varepsilon_\nu - \mu n_\nu)$], where $\beta = 1/k_BT$ and μ is the chemical potential.

$$f_F(\varepsilon_\nu) = (n_\nu) = \frac{\sum_{n_\nu=0,1} n_\nu e^{-\beta(n_\nu\varepsilon_\nu - \mu n_\nu)}}{\sum_{n_\nu=0,1} e^{-\beta(n_\nu\varepsilon_\nu - \mu n_\nu)}} = \frac{0 + e^{-\beta(\varepsilon_\nu - \mu)}}{1 + e^{-\beta(\varepsilon_\nu - \mu)}} = \frac{1}{e^{\beta(\varepsilon_\nu - \mu)} + 1}. \quad ...(55)$$

This is the famous Fermi–Dirac distribution, which we shall use later in the course.

Bosons: Wavefunctions and Occupation Number

Consider now two identical *bosons*, e.g. helium atoms, that are occupying two single-particle states $\psi_{\nu 1}$ and $\psi_{\nu 2}$. The two-particle wavefunction $\psi_{\nu_1\nu_2}(r_1, r_2)$ that explicitly fulfill the bosonic symmetry condition Eq. (51a) is

$$\psi_{\nu_1\nu_2}(r_1,r_2) = \frac{1}{\sqrt{2}}\left[\psi_{\nu_1}(r_1)\psi_{\nu_2}(r_2) + \psi_{\nu_1}(r_2)\psi_{\nu_2}(r_1)\right] \quad ...(56)$$

where the prefactor is to ensure the correct normalization:

$$\int dr_1 \int dr_2 \left|\psi_{\nu_1\nu_2}(r_1,r_2)\right|^2 = 1. \quad ...(57)$$

In contrast to the fermionic case there can be any number of bosons in a given orbital. In the case of N bosons occupying N orbitals (of which some may be the same) the operator $\hat{S}_+$ that symmetrizes the N single-particle wavefunction $\psi_{\nu_1}(r_1)\psi_{\nu_2}(r_2)\ldots\psi_{\nu_N}(r_N)$ can be written as "sign-less" determinant

$$\psi_{\nu_1\ldots\nu_N}(r_1,r_2\ldots,r_N) = \hat{S}_+\left[\psi_{\nu_1}(r_1)\psi_{\nu_2}(r_2)\ldots\psi_{\nu_N}(r_N)\right] \quad ...(58a)$$

$$= \frac{1}{\sqrt{N!}}\begin{vmatrix} \psi_{\nu_1}(r_1) & \psi_{\nu_1}(r_2) & \cdots & \psi_{\nu_1}(r_N) \\ \psi_{\nu_2}(r_1) & \psi_{\nu_2}(r_2) & \cdots & \psi_{\nu_2}(r_N) \\ \vdots & \vdots & \ddots & \vdots \\ \psi_{\nu_N}(r_1) & \psi_{\nu_N}(r_2) & \cdots & \psi_{\nu_N}(r_N) \end{vmatrix}_+ . \quad ...(58b)$$

The subscript " + " indicates that upon expanding this "*determinant*" all signs are taken to be plus.

Next we find the distribution function for bosons. Again using the grand canonical ensemble we derive the equally famous Bose–Einstein distribution $f_B(\varepsilon)$. It is derived like its fermionic counterpart, the Fermi–Dirac distribution $f_F(\varepsilon)$.

Consider a bosonic state characterized by its single-particle energy ε_ν. The occupation number of the state can be any non-negative integer $n_\nu = 0, 1, 2, \ldots$ In the grand canonical ensemble the average occupation number $\langle f_B(\varepsilon_\nu)\rangle$ is found by writing $\psi_\nu = e{-}\beta(\varepsilon_\nu - \mu)$ and using the formulas

$$\sum_{n=0}^{\infty} n\lambda^n = \lambda \frac{d}{d\lambda}\sum_{n=0}^{\infty}\lambda^n \text{ and } \sum_{n=0}^{\infty}\lambda^n = \frac{1}{1-\lambda}:$$

$$f_B(\varepsilon_\nu)=\frac{\sum_{n_\nu=0}^{\infty} n_\nu e^{-\beta(n_\nu\varepsilon_\nu-\mu n_\nu)}}{\sum_{n_\nu=0}^{\infty} e^{-\beta(n_\nu\varepsilon_\nu-\mu n_\nu)}}=\frac{\lambda_\nu \frac{d}{d\lambda_\nu}\sum_{n_\nu=0}^{\infty}\lambda_\nu^{n_\nu}}{\sum_{n_\nu=0}^{\infty}\lambda_\nu^{n_\nu}}=\frac{\frac{\lambda_\nu}{(1-\lambda_\nu)^2}}{\frac{1}{1-\lambda_\nu}}=\frac{1}{e^{\beta(\varepsilon_\nu-\mu)}-1}. \quad ...(59)$$

The *Bose–Einstein distribution* differs from the *Fermi–Dirac distribution* by having -1 in the denominator instead of +1. Both distributions converge towards the classical *Maxwell–Boltzmann distribution*, $f_M(n_\nu) = e^{-\beta(\varepsilon_\nu-\mu)}$, for very small occupation numbers, where the particular particle statistics is not felt very strongly.

Operators Acting on Many-particle States

The third fundamental postulate for many-particle systems is that the total action of few-particle operators are given by summation of their action on the individual coordinates. For a one-particle operator $\hat{O}_1(r_i)$ and a two-particle operator $\hat{O}_2(r_i, r_j)$ we have

$$O_1\psi=\sum_{i=1}^{N} O_1(r_i)\psi(r_1,r_2,\ldots r_N),$$

$$O_2\psi=\sum_{i>j}^{N} O_2(r_i,r_j)\psi(r_1,r_2,\ldots,r_N).$$

For fermions, when $\psi_{\nu_1\ldots\nu_N}(r_1,\ldots,r_N)=\hat{S}_-[\psi_{\nu_1}(r_1)\psi_{\nu_2}(r_2)\ldots\psi_{\nu_N}(r_N)]$ antisymmetrized product of single-particle orbitals, the single-particle operator acts evenly on all these single-particle orbitals, and we end up with the simple result

$$\langle O_1\rangle=\sum_{\nu_j}\langle O_1\rangle_{\psi\nu_j} \quad ...(61a)$$

$$=\sum_{\nu_j}\int dr\,\psi^*_{\nu_j}(r)\,O_1\psi_{\nu_j}(r). \quad ...(61b)$$

This is simply the sum of all the single-particle contributions.

Similarly for the two-particle operator O_2, which also for the same ψ acts evenly on all two-particle states $\psi_{\nu_j\nu_k}$ (the prefactor 1/2 takes care of double counting)

$$\langle O_2\rangle_\psi=+\frac{1}{2}\sum_{\nu_j\nu_k}\langle O_2\rangle_{\psi_{\nu_j\nu_k}} \quad ...(62a)$$

$$= +\frac{1}{2}\sum_{\nu_j \nu_k} \int dr_1 \int dr_2\, \psi^*_{\nu_j}(r_1)\psi^*_{\nu_k}(r_2)\, O_2 \psi_{\nu_j}(r_1)\psi_{\nu_k}(r_2) \quad \text{...(62b)}$$

$$-\frac{1}{2}\sum_{\nu_j \nu_k} \int dr_1 \int dr_2\, \psi^*_{\nu_j}(r_1)\psi^*_{\nu_k}(r_2) O_2 \psi_{\nu_j}(r_2)\psi_{\nu_k}(r_1). \quad \text{...(63c)}$$

Here we have used the explicit form Eq. (52) for the two-fermion state $\psi_{\nu_j \nu_k}(r_1, r_2)$. Note that the term where the *position vectors* r_1 and r_2 has been exchanged appears with a minus sign. This particular quantum expectation value has no classical analoge. It is intimately tied to the indistinguishability of quantum particles. This contribution to the expectation value is denoted the exchange term.

2

METALS AND CONDUCTION ELECTRONS

The study of electrons moving in a charge compensating background of positively charged ions is central in the understanding of solids. We shall restrict ourselves to the study of simple metals, but the results obtained are important for understanding of the *scanning tunnel microscope* (STM) and the *atomic force microscope* (AFM).

Any atom in a metal consists of three parts: the positively charged heavy nucleus at the center, the light cloud of the many negatively charged *core electrons* tightly bound to the nucleus, and finally, the outermost few *valence electrons*. The nucleus with its core electrons is called the *ion*. The ion mass is denoted M, and if the atom has Z valence electrons the charge of the ion is $+Ze$. To a large extend the inner degrees of freedom of the ions do not play a significant role leaving the center of mass coordinates R_j and total spin S_j of the ions as the only dynamical variables. In contrast to the core electrons the Z valence electrons, with mass m and charge $-e$, are often free to move away from their respective host atoms. Thus a gas of electrons swirling around among the ions is formed. In metals the electrons constituting this gas are able to conduct *electric currents*, and consequently they are denoted *conduction electrons*.

SINGLE-ELECTRON STATES: TRAVELLING WAVES

In our study of conduction electrons in metals we employ the particularly simple *Sommerfeld model*. In this model the ion charges are imagined to be smeared out to form a homogeneous and static

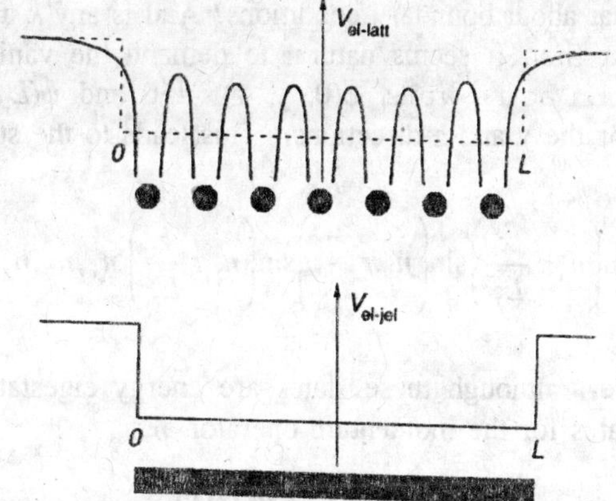

Fig. 2.1. A sketch showing the periodic potential, $V_{el\text{-}latt}$, present in a real lattice, and the imagined smeared out potential $V_{el\text{-}jel}$ of the jellium model.

positive charge density, $+Z\rho_{jel}$, called the *ion jellium*. Let the metal be shaped as a box with side lengths L_x, L_y, and L_z and volume $V = L_xL_yL_z$. The *periodic potential*, $V_{el\text{-}latt}$, present in a real lattice becomes the *constant potential* $V_{el\text{-}jel}$ confined by steep walls to the region $0 < x < L_x$, $0 < y < L_y$, and $0 < z < L_z$. We are free to choose the zero point of the energy, so it is natural to choose it such that the potential is zero, $V_{el\text{-}jel} = 0$, inside the box. In this case the Schrodinger equation Eq. (1) simply becomes

$$-\frac{\hbar^2}{2m}\nabla^2\psi(r) = E\psi(r). \qquad \text{...(1)}$$

It is straightforward to check that any plane wave with wavevector k,

$$\psi_k(r) = \frac{1}{\sqrt{V}}e^{ik\cdot r} = \frac{1}{V}e^{i(k_x x+k_y y+k_z z)}, \qquad \text{...(2)}$$

solves the Schrodinger equation with a k-dependent eigenenergy ε_k:

$$-\frac{\hbar^2}{2m}\nabla^2\psi_k(r) = \varepsilon_k\psi_k(r) \qquad \text{...(3a)}$$

$$\varepsilon_k = \frac{\hbar^2k^2}{2m} = \frac{\hbar^2}{2m}\left(k_x^2 + k_y^2 + k_z^2\right) \qquad \text{...(3b)}$$

We note that the wavefunctions are properly normalized since

$$\int dr|\psi(r)|^2 = \int dr\frac{1}{V} = 1.$$

But what about boundary conditions? And is any k an admissable solution? At first, it seems natural to demand the vanishing of the wave function at the walls, $\psi(0, y, z) = 0$ and $\psi(L_x, y, z) = 0$ (likewise for the y and z directions). This leads to the standing wave solutions

$$\widetilde{\psi}(r) = C \sin\left(n_x \pi \frac{x}{L_x}\right) \sin\left(n_y \pi \frac{y}{L_y}\right) \sin\left(n_z \pi \frac{z}{L_z}\right), n_x, n_y, n_z = 1,2,3,4,\ldots \qquad \ldots(4)$$

However, although these states are energy eigestates, they are not eigenstates for the momentum operator $\hat{p}$,

$$\frac{\hbar}{i} \nabla \widetilde{\psi}(r) \neq \hbar k \widetilde{\psi}(r). \qquad \ldots(5)$$

Instead we therefore choose the travelling wave solutions that arises from imposing periodic boundary conditions $\psi(L, y, z) = \psi(0, y, z)$ and $\psi'(L, y, z) = \psi'(0, y, z)$ (likewise for the y and z directions).

$$\psi_k(r) = \frac{1}{\sqrt{V}} e^{ik \cdot r}, \begin{cases} k_x = \frac{2\pi}{L_x} n_x \text{ (same for } y \text{ and } z) \\ n_x = 0, \pm 1, \pm 2, \ldots \text{ (same for } y \text{ and } z) \\ V = L_x L_y L_z, \end{cases} \qquad \ldots(6)$$

The travelling waves are eigenstates both to the energy operator $\hat{H}$ and to the momentum operator $\hat{p}$,

$$\hat{H} \psi_k(r) = \frac{\hbar^2 k^2}{2m} \psi_k(r) = \frac{h^2}{2m}\left(\frac{n_x^2}{L_x^2} + \frac{n_y^2}{L_y^2} + \frac{n_z^2}{L_z^2}\right) \psi_k(r) \qquad \ldots(7a)$$

$$\hat{p} \psi_k(r) = \frac{\hbar}{i} \nabla \psi_k(r) = \hbar k \, \psi_k(r). \qquad \ldots(7b)$$

Ground State for Non-interacting Electrons

Having identified the single-electron states we are now in a position to find the ground state for N *non-interacting electrons*. In this first analysis we neglect the *Coulomb interaction* between the electrons. We shall see later why this is justifiable.

For each allowed wavevector k, see Eq. (6), there exists an electron state ψ_k, that according to the *Pauli exclusion principle* can hold two electrons, one with spin up and one with spin down. The energy of an electron in this state is $\varepsilon_k = \frac{\hbar^2 k^2}{2m}$. It is natural to order the single-particle states according to their energies in ascending order,

$$\psi_{k_{1\uparrow}}, \psi_{k_{1\downarrow}}, \psi_{k_{2\uparrow}}, \psi_{k_{2\downarrow}}, \ldots \text{ where } \varepsilon_{k_1} \leq \varepsilon_{k_2} \leq \varepsilon_{k_3} \leq \ldots \qquad \ldots(8)$$

where the arrows indicate the electron spin in a given state. In the following we suppress this *spin index* and simply take spin into account by allowing two electrons (one with spin up and the other with spin down) in each state $\psi_k(r)$.

The *ground state*, i.e. the state with lowest energy, of a system containing N electrons is therefore constructed by filling up states with the lowest possible energy, i.e. the smallest values of k, until all N electrons are placed. If we represent the allowed k-vectors as points in a (k_x, k_y, k_z)-*coordinate system*, we easily see that the ground state is obtained by occupying with two electrons (spin up and spin down) all states inside the smallest sphere centered around the zero-energy state $k = 0$ containing $N/2$ allowed k-points.

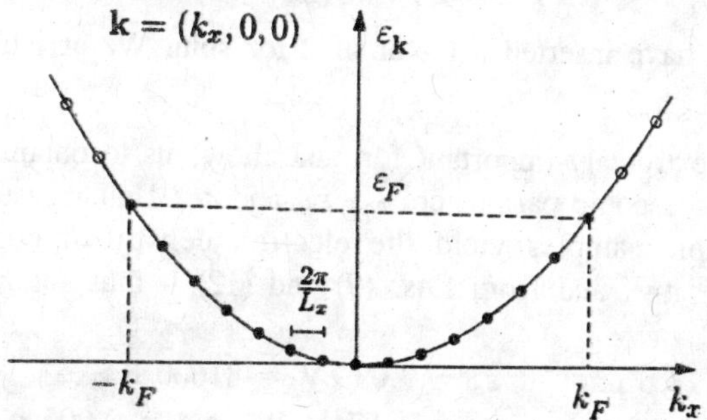

Fig. 2.2. The dispersion relation ε_k is plotted along the line $k = (k_x, 0,0)$, and ε_F and k_F are indicated.

This sphere in k-space representing the ground state of the non-interacting N-electron system is called the *Fermi sphere*. Due to the *Pauli principle* it is not possible to move electrons around deep inside the Fermi sphere, so when an N-electron system is disturbed by external perturbations most of the action involves occupied and unoccupied states near the surface. It is therefore relevant to focus on the surface. The radius of the Fermi sphere is denoted the Fermi wavenumber k_F. The energy of the top most occupied state is denoted the Fermi energy, ε_F. Associated with ε_F and the Fermi wavenumber k_F are the Fermi wave length λ_F and the Fermi velocity v_F:

$$\varepsilon_F = \frac{\hbar^2 k_F^2}{2m}, \; k_F = \frac{1}{\hbar}\sqrt{2m\varepsilon_F}, \; \lambda_F = \frac{2\pi}{k_F}, \; v_F = \frac{\hbar k_F}{m}. \qquad \ldots(9)$$

To facilitate calculations involving the Fermi sphere we note how the quantization of k in Eq. (6) means that one state fills a volume

$$\frac{2\pi}{L_x}\frac{2\pi}{L_y}\frac{2\pi}{L_z} = \frac{(2\pi)^3}{V}$$

in k-space. From this follows the very useful rule of how to evaluate k-sums by k-integrals:

$$\sum_k \rightarrow \frac{V}{(2\pi)^3}\int dk. \qquad \text{...(10)}$$

We shall immediately use this rule to calculate the relation between the macroscopic quantity $n = N/V$, the density, and the microscopic quantity k_F, the Fermi wavenumber.

$$N = \sum_{\text{allowed } k}^{k_F} 2 = 2\frac{V}{(2\pi)^3}\int_{|k|<k_F} dk = 2\frac{V}{(2\pi)^3}\int_0^{k_F} 4\pi k^2\, dk = \frac{V}{3\pi^2}k_F^3, \qquad \text{...(11)}$$

where we have inserted a factor of 2 for spin. We arrive at

$$k_F^3 = 3\pi^2 n. \qquad \text{...(12)}$$

This extremely important formula allows us to obtain the values of the microscopic parameters k_F, ε_F, and υ_F. Hall measurements in macroscopic samples yield the electron density of copper, $n = 8.47\times10^{28}$ m^{-3}, and from Eqs. (9) and (12) it thus follows that for copper

$$\begin{aligned} k_F &= 13.6 \text{ nm}^{-1} & \varepsilon_F &= 7.03 \text{ eV} = 81600 \text{ K} \\ \lambda_F &= 0.46 \text{ nm} & \upsilon_F &= 1.57\times10^6 \text{ m/s} = 0.005\, c. \end{aligned} \qquad \text{...(13)}$$

Note that the Fermi energy corresponds to an extremely high temperature, which we shall return to shortly, and even though the Fermi velocity is large it is still less than a percent of the velocity of light, and we need not invoke relativistic considerations.

Energy of the Non-Interacting Electron Gas

In this section we calculate the ground state energy of the electron gas in the limit of high electron densities. The reason for choosing this limit is, as we shall see, that the Fermi sphere is an excellent approximation to the ground state even if the Coulomb interaction between the electrons is taken into account. The question is why this is so. The key to the answer lies in how the kinetic energy E_{kin} and the potential Coulomb energy E_{pot} of a typical electron depend on the electron density n. From Eqs. (9) and (12) we have $E_{\text{kin}} \approx \varepsilon_F \propto k_F^2 \propto n^{2/3}$. At the density n the typical distance d between the electrons is proportional to $n^{-1/3}$, and consequently the Coulomb energy behaves as $E_{\text{pot}} \propto d^{-1} \propto n^{1/3}$. So we find that

$$\frac{E_{\text{port}}}{E_{\text{kin}}} \propto \frac{n^{\frac{1}{3}}}{n^{\frac{2}{3}}} = n^{-\frac{1}{3}} \underset{n\to\infty}{\to} 0, \qquad \text{...(14)}$$

revealing the following perhaps somewhat counter intuitive fact: the importance of the electron-electron interaction diminishes as the density of the electron gas increases. Due to the Pauli exclusion principle the kinetic energy simply becomes the dominant energy scale in the interacting electron gas at high densities.

We move on to calculate the ground state energy for the non-interacting electron gas. Due to the absense of interaction this is purely kinetic energy:

$$E_{\text{kin}} = \sum_{\text{allowed}\,k}^{k_F} 2\,e_k = \frac{\hbar^2}{2m}\frac{2V}{(2\pi)^3}\int_{|k|<k_F} dk\, k^2 = \frac{V\hbar^2}{m(2\pi)^3}\int_0^{k_F} 4\pi k^4\, dk = \frac{3}{5}N\varepsilon_F. \qquad \text{...(15)}$$

The kinetic energy per particle as a function of the density n is therefore

$$\frac{E_{\text{kin}}}{N} = \frac{3}{5}\frac{\hbar^2}{2m}k_F^2 = \frac{3}{5}\frac{\hbar^2}{2m}(3\pi^2)^{\frac{2}{3}}n^{\frac{2}{3}}. \qquad \text{...(16)}$$

ENERGY OF THE INTERACTING ELECTRON GAS

The Coulomb interaction is now taken into account. First we note that the sum of classical electrostatic energies from ion-ion, ion-electron, and electron-electron interactions is zero. This is because the both the electrons and the ions are smeared out homogeneously leaving a system that is completely charge neutral everywhere in space. Any non-zero contribution to the Coulomb energy must therefore be of quantum nature.

In quantum theory the Coulomb interaction is given by a two-particle operator $\hat{V}_{\text{Coul}}$, since two particles are involved each with its own coordinates,

$$\hat{V}_{\text{Coul}}(r_1, r_2) = \frac{e^2}{4\pi\varepsilon_0}\frac{1}{r_2 - r_1}. \qquad \text{...(17)}$$

We disregard the spin index here. The interesting quantum effects in the following comes only from electron pairs with the same spin. The potential energy E_{pot} is just the expectation value of the Coulomb interaction operator $\hat{V}_{\text{Coul}}$ acting on the N-particle state ψ, the Fermi sphere. Using the results in particular the equation valid for two-particle operators we get

$$E_{\text{pot}} = \left\langle \hat{V}_{\text{Coul}} \right\rangle_{\psi} \qquad \text{...(18a)}$$

$$= +\frac{1}{2}\sum_{k_1k_2}\left\langle \hat{V}_{\text{Coul}} \right\rangle_{\psi_{k_1k_2}} \qquad \text{...(18b)}$$

$$= +\frac{e^2}{8\pi\varepsilon_0}\sum_{k_1k_2}\int dr_1 \int dr_2 \frac{\left|\psi_{k_1}(r_1)\right|^2\left|\psi_{k_2}(r_2)\right|^2}{|r_2 - r_1|} \qquad \text{...(18c)}$$

$$-\frac{e^2}{8\pi\varepsilon_0}\sum_{k_1k_2}\int dr_1 \int dr_2 \frac{\psi^*_{k_1}(r_1)\psi^*_{k_2}(r_2)\psi_{k_1}(r_2)\psi_{k_2}(r_1)}{|r_2 - r_1|}. \qquad \text{...(18d)}$$

The term Eq. (18c) is easy to interpret. It is simply the classical electrostatic energy from two electron charge distributions $|\psi_{k_1}(r_1)|^2$ and $|\psi_{k_2}(r_2)|^2$. We have already discussed how this is cancelled by the interaction with the positive ions. However, the term Eq. (18d) is not a classical electrostatic energy, and it will therefore yield a *non-zero contribution*. We note that it involves exchange of coordinates between the two single-particle orbitals ψ_{k_1} and ψ_{k_2} typical of quantum states with two (or more) indistinguishable particles. The term is therefore called the *exchange energy*, denoted E_{exch}. It is not difficult to calculate the integral when ψ_k are travelling waves, however, it is somewhat cumbersome. We therefore just state the result:

$$E_{\text{exch}} = \frac{-e^2}{4\pi\varepsilon_0 V^2}\sum_{k_1k_2}^{k_F}\int dr_1 \int dr_2 \frac{e^{i(k_1-k_2)\cdot(r_2-r_1)}}{|r_2 - r_1|} = -\frac{3N}{4\pi}\frac{e^2}{4\pi\varepsilon_0}k_F. \qquad \text{...(19)}$$

Or expressing k_F in terms of density:

$$\frac{E_{\text{exch}}}{N} = -\frac{3(3\pi^2)^{\frac{1}{3}}}{4\pi}\frac{e^2}{4\pi\varepsilon_0}n^{\frac{1}{3}}. \qquad \text{...(20)}$$

We note that the exchange energy E_{exch} is negative. This can be explained by noting that the Pauli exclusion principle prohibits electrons to be close one another. This quantum effect therefore forces the electrons to stay further apart on average than would have been the case in a purely classical world. But larger distance means smaller *Coulomb energy*, and the quantum correction to the classical Coulomb energy must therefore be negative.

The total energy per electron E/N in the metal is now obtained by combining the quantum kinetic energy Eq. (16) with the quantum exchange energy Eq. (20):

$$\frac{E}{N}=\frac{E_{kin}+E_{exch}}{N}=\frac{3}{5}\frac{\hbar^2}{2m}(3\pi^2)^{\frac{2}{3}}n^{\frac{2}{3}}-\frac{3(3\pi^2)^{\frac{1}{3}}}{4\pi}\frac{e^2}{4\pi\varepsilon_0}n^{\frac{1}{3}}. \quad ...(21)$$

It is common practice to express the result in terms of a dimensionless measure r_s of the average distance between the electrons and the *Bohr radius* $a_0 = 0.053$ nm. By definition $r_s a_0$ is the radius of a sphere containing exactly one electron,

$$\frac{4\pi}{3}(r_s a_0)^3=\frac{1}{n}\Rightarrow r_s=\left(\frac{3}{4\pi}\right)^{\frac{1}{3}}\frac{1}{n^{\frac{1}{3}}}\frac{1}{a_0}. \quad ...(22)$$

Putting in numbers we obtain for the total energy per particle:

$$\frac{E}{N}=\left(\frac{2.211}{r_s^2}-\frac{0.916}{r_s}\right)13.6\,eV. \quad ...(23)$$

This result shows that the electron gas is stable when the repulsive Coulomb interaction is turned on. No external confinement potential is needed to hold the electron gas in the ion jellium together. There exists an optimal density n^*, or inter-particle distance r_s^*, which minimizes the energy and furthermore yields an energy $E^* < 0$. The negative exchange energy overcomes the positive kinetic energy. The equilibrium situation is obtained from $\frac{\partial}{\partial r_s}(E^{(0)}+E^{(1)})=0$, and we can compare the result with experiment:

$$r_s^* = 4.83,\ \frac{E^*}{N} = -1.29\ eV \text{ (the jellium mode)} \quad ...(24a)$$

$$r_s = 3.96,\ \frac{E}{N} = -1.13 \text{ eV (experiment on Na)} \quad ...(24b)$$

It is gratifying that our crude jellium model is able to grasp some central features in the physics of metals. It shows how kinetic and

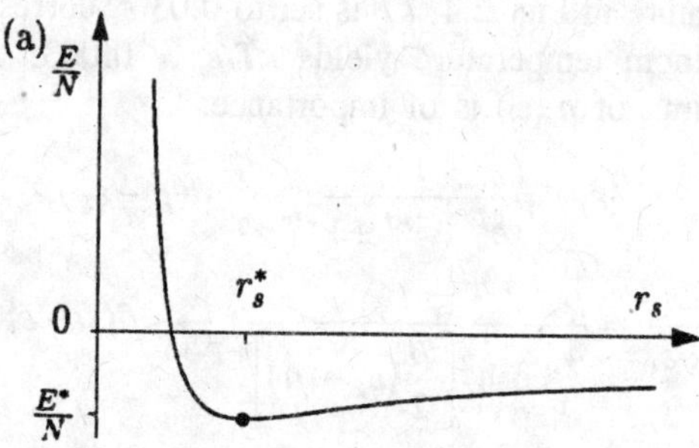

Fig. 2.3. The energy per particle E/N of the 3D electron gas in the jellium model as a function of the dimensionless inter-particle distance r_s.

potential energies of quantum nature keep the metal together, and it predicts an almost correct electron density and ionization energy.

DENSITY OF STATES

The concept of density of states, $D(\varepsilon) = \frac{dN}{d\varepsilon}$, is widely used. The function $D(\varepsilon)$ simply gives the number ΔN of states in the energy interval $\Delta\varepsilon$ around the energy ε, $\Delta N = D(\varepsilon)\Delta\varepsilon$, and the density of states per volume $d(\varepsilon) = D(\varepsilon)/V = \frac{dn}{d\varepsilon}$. Again using Eq. (12) we find

$$\varepsilon_F = \frac{\hbar^2}{2m}k_F^2 = \frac{\hbar^2}{2m}(3\pi^2)^{\frac{2}{3}}n^{\frac{2}{3}} \Rightarrow n(\varepsilon) = \frac{1}{3\pi^2}\left(\frac{2m}{\hbar^2}\right)^{\frac{3}{2}}\varepsilon^{\frac{3}{2}}, \text{ for } \varepsilon > 0, \quad ...(25)$$

and from this

$$d(\varepsilon) = \frac{dn}{d\varepsilon} = \frac{1}{2\pi^2}\left(\frac{2m}{\hbar^2}\right)^{\frac{3}{2}}\varepsilon^{\frac{1}{2}}\theta(\varepsilon), \quad D(\varepsilon) = \frac{dN}{d\varepsilon} = \frac{V}{2\pi^2}\left(\frac{2m}{\hbar^2}\right)^{\frac{3}{2}}\varepsilon^{\frac{1}{2}}\theta(\varepsilon) \quad ...(26)$$

Here we have introduced the much used Heavy side step function $\theta(\varepsilon)$

$$\theta(\varepsilon) = \begin{cases} 1, \text{ for } \varepsilon > 0, \\ 0, \text{ for } \varepsilon < 0. \end{cases} \quad ...(27)$$

The density of states $D(\varepsilon)$ is a very useful function. It can for example be used to calculate the particle number, $N = \int d\varepsilon\, D(\varepsilon)$, and the total energy, $E^{(0)} = \int d\varepsilon\, \varepsilon\, D(\varepsilon)$.

ELECTRON GAS AT FINITE TEMPERATURE

In the previous section temperature was tacitly taken to be zero. As temperature is raised from zero the occupation number is given by the Fermi-Dirac distribution $n_F(\varepsilon_k)$. The main characteristics of this function is shown in Fig. 2.4. Note that to be able to see any effects of the temperature in Fig. 2.4, kT is set to 0.03 ε_F corresponding to $T \approx 2400$ K. Room temperature yields $kT/\varepsilon_F \approx 0.003$, thus the low temperature limit of $n_F(\varepsilon_k)$ is of importance:

$$n_F(\varepsilon_k) = \frac{1}{e^{\beta(\varepsilon_k - \mu)} + 1} \underset{T\to 0}{\rightarrow} \theta(\mu - \varepsilon_k), \quad ...(28a)$$

$$-\frac{\partial n_F}{\partial \varepsilon_k} = \frac{\beta}{4}\frac{1}{\cosh^2\left[\frac{\beta}{2}(\varepsilon_k - \mu)\right]} \underset{T\to 0}{\rightarrow} \delta(\mu - \varepsilon_k). \quad ...(28b)$$

At $T = 0$ the chemical potential μ is identical to ε_F. But in fact μ varies slightly with temperature. A careful analysis based on the so-

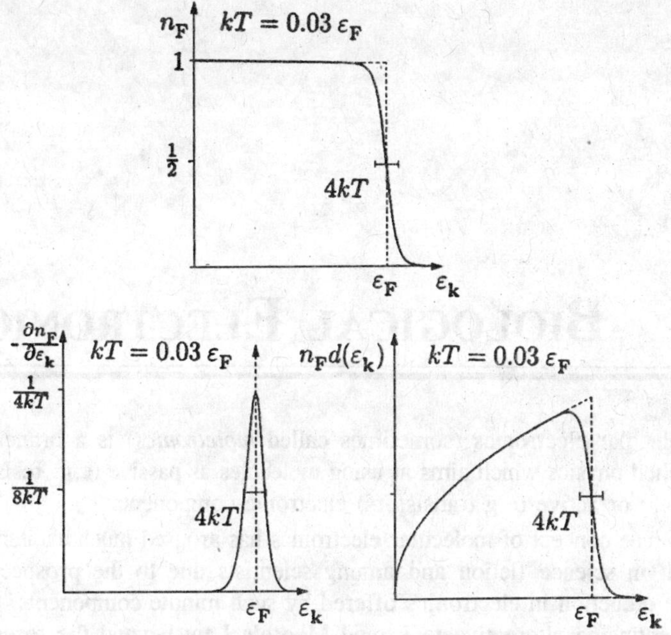

Fig. 2.4. The Fermi-Dirac distribution.

called *Sommerfeld expansion* combined with the fact that the number of electrons does not change with temperature yields

$$n(T=0) = n(T) = \int_0^\infty d\varepsilon\, d(\varepsilon) f(\varepsilon) \Rightarrow \mu(T) = \varepsilon_F \left[1 - \frac{\pi^2}{12}\left(\frac{kT}{\varepsilon_F}\right)^2 + \ldots\right] \quad \ldots(29)$$

Because ε_F according to Eq. (13) is around 80000 K for metals, we find that even at the melting temperature of metals only a very limited number ΔN of electrons are affected by thermal fluctuations. Indeed, only the states within $2kT$ of ε_F are actually affected, and more precisely we have $\Delta N/N = 6kT/\varepsilon_F$ ($\approx 10^{-3}$ at room temperature). The Fermi sphere is not destroyed by heating, it is only slightly smeared. Now we have at hand an explanation of the old paradox in thermodynamics, as to why only the ionic vibrational degrees of freedom contribute significantly to the specific heat of solids. The electronic degrees of freedom are simply 'frozen' in. Only at temperatures comparable to ε_F they begin to play a major role.

3

BIOLOGICAL ELECTRONICS

Molecular electronics (sometimes called *moletronics*) is a branch of applied physics which aims at using molecules as passive (e.g. resistive wires) or active (e.g transistors) electronic components.

The concept of molecular electronics has aroused much excitement both in science fiction and among scientists due to the prospect of size reduction in electronics offered by such minute components. It is an entincing alternative to extend Moore's Law beyond the foreseen limits of small-scale conventional silicon integrated circuits. As a result, molecular electronics is currently a very active research field, sometimes marked by controversy since many fundamentals questions concerning both theory and experiments are left open.

Among the important issues is the determination of the resistance of a single molecule (both theoretical and experimental). Another problem faced by this field is the difficulty to perform direct characterization since imaging at the molecular scale is often impossible in many experimental devices.

Study of charge transfer in molecules was advanced in the 1940s by Robert Mulliken and Albert Szent-Gyorgi in discussion of so-called "donor-acceptor" systems and developed the study of charge transfer and energy transfer in molecules. Likewise, a 1974 paper from Mark Ratner and Avi Aviram illustrated a theoretical molecular rectifier. Later, Aviram detailed a single-molecule field-effect transistor in 1988. Further concepts were proposed by Forrest Carter of the Naval Research Laboratory, including single-molecule logic gates.

Apart from the Aviram and Ratner proposal, molecular electronics received an initial boost from the experimental discovery of conducting polymers in the mid-seventies. Before this date, organic molecules (which form crystals or polymers) were considered insulating or at best weakly conducting semiconductors. McGinness, Corry, and Proctor reported the first molecular electronic device (a voltage-controlled bistable switch) in Science magazine. As its active element, this device used melanin, a mixed polymer of polyacetylene, polypyrrole, and polyaniline. This device is now in the Smithsonian's collection of historic electronic devices. As Hush notes, their material also showed negative differential resistance, "a hallmark of modern advances in molecular electronics". Polymeric materials are solids made of molecular chains (macromolecules) which are bound to each other via weak electrostatic interactions. Each chain is a regular repetition of a small ensemble of atoms called a monomer. In polyacetylene, the monomer is made of three carbon atoms linked alternatively by a single sigma bond and a double bound. Because of this alternate structure, the polymer is said to be conjugated. In conjugated structures, electrons are more delocalized than in other organic structures, which a-posteriori explains the experimental properties found by Me Giness et al.

However, conjugation is not sufficient to obtain useful conduction from polymers. A few years later, in 1977, through the introduction of dopants (such as halogen atoms), chemists started to greatly improve the conductance of conjugated polymers. These findings opened the door to plastic electronics and optoelectronics which are beginning to find extensive commercial application.

In polymers, classical organic molecules are composed of both carbon and hydrogen (and sometimes additional compounds such as nitrogen, chlorine or sulfur). They are obtained from petrol and can often be synthesized in large amounts. Most of these molecules are insulating when their length exceeds a few nanometers. However, naturally occurring carbon is conducting. In particular, graphite (recovered from coal or encountered naturally) is conducting. From a theoretical point of view, graphite is a semi-metal, a category in between metals and semiconductors. It has a layered structure, each sheet being one atom thick. Between each sheet, the interactions are weak enough to allow an easy manual cleavage.

Tailoring the graphite sheet to obtain well defined nanometer-sized objects remains a challenge. However, by the close of the twentieth century, chemists were exploring methods to fabricate extremely small graphitic objects that could be considered single molecules. After studying the interstellar conditions under which carbon is known to form clusters, Richard Smalley's group set up an experiment in which graphite was vaporized using laser irradiation. Mass spectrometry revealed that clusters containing specific "magic numbers" of atoms were stable, in particular those clusters of 60 atoms. Harry Kroto, an English chemist who assisted in the experiment, suggested a possible geometry for these clusters—atoms covalently bound with the exact symmetry of a soccer ball. Coined buckminsterfullerenes, buckyballs or C60, the clusters retained some properties of graphite, such as conductivity. These objects were rapidly envisioned as possible building blocks for molecular electronics.

Recent progress in nanotechnology and nanoscience has facilitated both experimental and theoretical study of molecular electronics. In particular, the development of the scanning tunneling microscope (STM) and later the atomic force microscope (AFM) have facilitated manipulation of single-molecule electronics.

A collaboration of researchers at HP and UCLA, led by James Heath, Fraser Stoddart, R. Stanley Williams, and Philip Kuekes, has developed molecular electronics based on rotaxanes and catenanes.

Work is also being done on the use of single-wall carbon nanotubes as field-effect transistors. Most of this work is being done by IBM.

The Aviram-Ratner model for a molecular rectifier, which until recently was entirely theoretical, has been confirmed experimentally and unambiguously in a number of experiments by a group led by Geoffrey J. Ashwell at Cranfield University, UK. Many rectifying molecules have so far been identified, and the number and efficiency of these systems is expanding rapidly.

Supramolecular electronics is a new field that tackles electronics at a supramolecular level.

Chemoelectric Cells

The human body provides a renewable source of chemoelectric energy that may be tapped by various means. As long ago as 1959, Pinneo and Kesselman reported powering an FM transmitter by simply inserting two steel electrodes into the brain of a cat, generating 0.5 microamperes at 40 millivolts. In another experiment by Reynolds in 1963, electrodes inserted subcutaneously and abdominally into an anesthetized rat produced 400 millivolts and ~ 10 microwatts, powering an oscillator circuit for 8 hours with no sign of decline. Since then, three classes of bioelectric energy have been distinguished: ionic concentration, biogalvanic, and biofuel cells.

Interestingly, when the rat was finally administered an instantly lethal intracardial injection of Nembutal the voltage required 75 minutes, after death, to fall to zero.

The ionic concentration cell may exploit differences in chemical concentration by placing two similar electrodes in different compartments of the body having different chemical composition, generating a voltage proportional to the logarithm of the concentration ratio. By this means, it has been proposed that power could be drawn from the differences in oxygen and hydrogen ion concentration between arterial and venous blood, or the acid-alkali differences between the stomach fluids and surrounding tissue. Similarly, transcytomembrane electrodes could develop potentials of 1-100 mV between cytosol and intercellular fluid. This would allow a 10 pW nanorobot to be powered by a current flow of 10 nanoamperes at 1 mV through a 10^5 ohm load resistance, although power draws exceeding ~100 pW might prove electrokinetically disruptive to the cell. Also, this source may be less reliable for dying or challenged human tissue.

Biogalvanic energy sources exploit the electrochemical potential between metallic electrodes (the fuel) in an electrolyte solution. Two broad classes of galvanic pairs which generate electrical energy have been investigated for medical use: those in which both anode and cathode are consumed and go into solution, and those in which the cathode is inert. In the first group, zinc (anode) and silver chloride (cathode) electrodes have been implanted in human test subjects for up to two years. The Zn goes into solution as Zn^{++}; the AgCl cathode is converted into Ag and the Cl ion goes into solution. This chemoelectric source generates steady currents of ~0.2 picoamperes/micron at ~1 volt or ~10 pW/50 microns of electrode surface at 70%-90% efficiency, an energy density of 10^7 watts/m assuming 20-nm thick electrodes. Similar output is obtained from galvanic pairs in the second group, of which the best studied is a zinc anode coupled with a palladium or platinum-black cathode; molecular oxygen combines with water at the cathode producing OH and H^+ ions. Unfortunately, galvanic sources provoke a significant host reaction, including formation of layers of necrotic debris, free neutrophils, granulation tissue and complete fibrous connective tissue encapsulation of long-term implants. The biofuel cell relies upon redox reactions in which neither anode nor cathode is consumed but merely act as catalysts, potentially a great advantage in nanomedical applications. The biofuel cell that has received the most attention for nanomedicine is the oxyglucose cell, which could rely upon an electrochemical process in which glucose is oxidized at the anode and molecular oxygen is reduced at the cathode.

Decades-old experiments using platinum black electrodes produced ~0.6 picoamps per micron of electrode surface at ~0.3 volts, with power levels of 0.2 pW/micron in vitro (declining over time due to the formation and absorption of gluconic acid at the anode), but only 0.004 pW/micron in vivo (when implanted in rat or rabbit test animals) in part due to poisoning of the catalytic anode action by proteins. Complete electrochemical oxidation of glucose has not yet been demonstrated experimentally. Controlled-permeability membranes may eliminate these problems in nanomedical applications, but even using the aforementioned electrodes might allow a continuous power density of 10^7 watts/m for a 10 pW oxyglucose biofuel cell using 50 micron of electrodes 20-nm thick. Since gluconic acid can also be oxidized at potentials which oxidize glucose, gluconate may not be the only by-product, thus appropriate means of disposing of any potentially undesirable by-products must be included in a practical design. The addition of specific catalysts (e.g., ruthenium) to the electrodes could be helpful.

Ethanol fuel cells, though inefficient, are already in wide commercial use. For example, the Lion alcolmeter fuel cell alcohol sensor employs catalytic platinum electrodes (supported on a ~1 cm PVC matrix disk assembly impregnated with acidic electrolyte) to produce electron flow by oxidizing ethanol to acetic acid (producing two free electrons per alcohol molecule) at the anode and reducing atmospheric oxygen at the cathode. Normal alcohol working concentrations are 5900 parts per million typically producing up to ~10 microamps at ~5 millivolts across a 390 ohm load, or ~50 nanowatts (power density at most ~0.1 watt/m). The Lion fuel cell can also produce energy from all primary and secondary aliphatic alcohols but not from aldehydes, ketones, ether, esters, hydrocarbons or carboxylic acids. Other manufacturers of electrochemical fuel cells for alcohol sensing include PAS Systems, Guth Laboratories, and Intoximeter. Direct methanol fuel cells also were under development in 1998.

The following analysis was prompted by R. Merkle's suggestion that an oxyglucose biofuel cell using nanoscale membranes might achieve significantly higher power densities than indicated by past experiments. Consider a proton (H^+) exchange nanomembrane consisting of a 1 nm thick diamondoid sheet containing a number of very narrow pores each lined with atoms of oxygen, fluorine or nitrogen, creating negatively charged channels with high proton affinity. A channel narrow enough to admit only protons but nothing else (~0.1 nm, excluding even helium atoms) and surrounded by a ~3 atom thick support wall makes a ~1 nm pore structure ~1 nm wide.

The power generated by each pore due to proton flow is P_{pore} = q 'N V_p, taking q = 1.6 × 10^{-19} joule/proton-volt, 'N is the proton flow rate in protons/sec, and V_p is the electrical potential through which the charges fall. Experimental observations suggest V_p may reach 0.75 volt at zero load, but averages 0.3-0.6 volts at moderate loads; for this analysis, V_p will be taken as ~0.5 volt. As for 'N, the nicotinic acetylcholine receptor channel has a 0.65-2.2 nm inside diameter and allows 2.5 × 10^7 Na^+ ions/sec to pass while the channel is open, comparable to the transport rate of artificial transmembrane peptide nanotube ion channels. Taking this flow rate as representative for 'N, then P_{pore} ~ 2 pW/pore.

Redox fuel cells using transition metal catalysts typically achieve n_{cat} ~10 catalytic events/sec per catalytic atom. Oxidation of each glucose molecule produces 24 protons, so the required catalytic rate is 'N/24 ~ 10^6 glucose molecules/ sec per pore which in turn requires 'N / (24 n_{cat}) ~ 10^5 metal catalyst atoms per pore. Taking Pt atoms as representative, 10^5 Pt catalyst atoms have a volume of ~1500 nm. Including the 1 nm pore structure and doubling the volume to allow for catalyst dispersal, fluid access and structural overhead gives a 3000 nm single-pore catalytic unit generating 2 pW, or a power density of ~7 × 10^{11} watts/ m. (Assumed catalyst atom usage is ~10^{17} atoms/watt; in 1999, the best Pt-catalyzed proton exchange membranes required ~10^{18} atoms/watt.) The current flow of 4 picoamps through each (0.1 nm) pore gives a current density of 4 × 10^8 amps/m, well below the ~10^{10} amp/m maximum current density in bulk aluminum. Since ~10^6 glucose molecules/sec are consumed per pore and each glucose molecule represents 4765 zJ of free energy, then total energy available is ~4.8 pW/pore. Thus it appears the device may be 2 pW/ 4.8 pW ~40% efficient, though this figure is highly voltage-dependent.

J. Soreff notes that the oxyglucose biofuel cell can in theory operate near the thermodynamic limit but may require the design of a complex sequence of catalytic reactions, not unlike microbial metabolism. The glucose/gluconic acid source may have lower efficiency due to incomplete oxidation but could be useful in early systems because it requires optimization of catalytic sites for just one reaction. On the other hand, the production of protons and electrons from complete oxidation of glucose potentially encompasses a lengthy biochemical chain which possibly may be implemented using sets of immobilized enzymes in the manner of mitochondrial respiration. (Mitochondrial power density is 10^5-10^6 watts/m.)

There are limits to efficiency set by mismatches between redox potentials at various steps in a respiratory chain. Given a single anodic chamber with a common electric potential and pH, concentrations of intermediates must be unequal to compensate for the redox potential mismatches at various steps. If these concentrations are too unequal, then the lowest concentration in the chain becomes the rate-limiting intermediate for the entire process. Throughput may be optimized by careful choice of intermediates, couplings between reactions, and catalyst design. To oxidize glucose, in biology, glycolysis requires ~18 enzymes, the TCA cycle ~9 enzymes, and the pentose phosphate pathway ~9 enzymes. Soreff suggests that a slightly different strategy is to have separate oxidation/proton chambers for each stage in the catalytic pipeline, allowing intermediates to diffuse through semipermeable membranes or to be transported via molecular sorting rotors. This transfers the problem of matching redox potentials into the electrical domain where it is more easily resolved.

Oxyhydrogen fuel cells are another possibility, though probably at much lower power density because hydrogen must first be produced from the glucose fuel. This occurs naturally in the energy-generating hydrogenosomes of the trichomonads (which use the enzymes pyruvate:ferredoxin oxidoreductase and hydrogenase) and in the laboratory using related enzymes (from bacteria that live near hot underwater vents) to convert glucose into hydrogen gas and water. In a modern 350 K oxyhydrogen fuel cell, O_2 and H_2 are fed into adjacent chambers separated by a proton exchange membrane. This membrane allows only H^+ ions to flow to the O_2 side, making water as the sole waste product and establishing a net negative charge on the H_2 anode and a net positive charge on the O_2 cathode, an electrical potential of ~0.6 volts under load. This chemoelectric process is at least ~50% efficient but typically achieves only 10^4-10^6 watts/ m partly due to hydrogen's low volumetric energy density and partly due to the crude membranes currently in commercial use. Bacterium-based biofuel cells have also been investigated.

Electrical Energy Conversion Processes

Microscale electrostatic motors have been fabricated since the late 1980s and are the subject of great current research interest.

Commercially available micromachined rotors ~1 micron thick and ~100 microns in radius respond to electric field intensities $>10^8$ volts/m producing motive torques of ~10 pN-m by converting electrical to mechanical energy, achieving power densities in the 10^4-10^5 watts/m range with device lifetimes approaching 10^7 rotations.

Drexler has described a class of submicron direct-current (DC) electrostatic motors capable of converting between mechanical and electrical power in either direction at a power density of $\sim 4 \times 10^{14}$ watts/m and an estimated efficiency >99%. (Typical macroscale brush DC motors can have 55%-75% efficiencies, while DC brushless motors can have efficiencies as high as 95%.)

In one implementation, electric charge is placed on the rim of a rotor as the rim passes within a dee electrode. This charge is then transported across a ~3 nm gap to the interior of the opposite dee electrode, where it is removed and replaced by a charge of opposite sign. Applying a voltage of proper sign across the dees causes the charges in transit to apply a torque to the rotor (like a Van de Graff generator operating in reverse), converting electrical into mechanical power. As a specific example, a motor 390 nm in diameter and 25 nm thick driven by a electric field strength of $\sim 0.2 \times 10^9$ volts/m spins with a near-maximum rim velocity of ~1000 m/sec; a current of ~110 nanoamperes and an electric potential of ~10 volts yields a power output of ~1.1 microwatts. Single electron-transfer events at nanometer electrodes have been generated and measured experimentally, and macroscale "Dirod" electrostatic generators have been operated as electrostatic motors in a student laboratory.

Electrical energy is also converted into mechanical energy via the inverse piezoelectric effect. Piezoelectric materials such as lead-zirconate-titanate ceramics (PZT) and natural mechanical resonators such as quartz, when subjected to an electric field, expand or contract depending on the orientation of the field to the piezoelectric polarization. For example, applying 100 volts across a 750 micron thick piezoelectric crystal causes it to deform by ~37 nm, or ~0.4 nm/volt. There are also piezoelectric rotary actuators, with rotations measured in ~arcsec. The feasibility of fabricating ~1 micron thick quartz resonators by etching techniques was demonstrated decades ago. PZT crystals are commonly used to provide movement of samples being examined in scanning probe microscopes such as STMs and AFMs. Piezoelectric drivers used in bimorph configurations as electromechanical transducers can produce large mechanical displacements but relatively small forces.

Dielectric induction micromotors 2 microns thick and 100 microns in diameter operating in water can be driven at 250 rev/sec at 110 volts applied potential, converting electrical to mechanical motion and producing a torque of 0.3 pN-m. Electroactive polymers (EAPs) and electrostrictive polymers such as iPMMA are being investigated for use in "artificial muscles"; an exceptionally high electrostrictive response (~4%) has been observed in electron-irradiated P(VDF-TrFE) copolymer. Electromechanical actuators based on sheets of single-walled carbon nanotubes have been described. Even bone can be made into an electret.

Electrostatic actuators can produce substantial mechanical forces. One electromechanical transducer described by Drexler is a capacitor with one plate fixed and connected to a signal source, with the other plate grounded via a tunneling junction, leaving it free to move within a small range of displacements. One design with a 1-nm stroke length and a (12 nm) plate area transduces an electrical voltage varying from 0-5 volts into a mechanical signal ranging from 0-1 nN.

Electrothermal energy conversion occurs during simple joule heating, or during Seebeck effect or Peltier effect cooling, when current passes through a bimetallic junction causing one side to get hot and the other side cold.

Electrochemical transduction may occur in enzymes with electric field-sensitive conformational states, in voltage-gated ion channels and nanomembranes, and in electrolytic cells which are chemoelectric cells operated in reverse.

Quantum well nanostructures can be useful in fabricating electrooptical transducers such as light-emitting diodes, electroluminescent displays and submicron-scale solid-state lasers, and electrochromic glass with voltage-dependent transmittance permits optical gating. Electric current sent through carbon nanotubes causes both field emission and luminescence, about one photon per 10^6 electrons —even a single nanotube makes a faint but visible glow, another case of electrooptical energy transduction.

Magnetomechanical transduction has also been employed to remotely power centimeter-size simple legged robots using magnetostrictive alloys as magnetic field-driven actuators in an 80 Hz 1-tesla field, but this approach may find only limited utility in nanomedical systems because molecular-scale drives suffer from the adverse scaling properties of electromagnet fields. Magnetic sensors, perhaps using molecular magnets, possibly could be employed as power transducers.

Electrical Tethers

One of the most obvious methods for providing energy to nanodevices is simply to attach a power cord to each machine. The bending moment of a wire is proportional to its radius to the fourth power, so molecular nanowires are much more flexible than macroscopic wires. These power cables could be made of traditional metallic conductors (15-nm thick gold wires have been fabricated on a Si substrate) or doped polyacetylene chains and other organic conductors which have been used to make LEDs and polymer batteries.

Wires with electrical and thermal insulation can safely carry higher currents in vivo. A 100-micron thick water jacket allows electrical current and voltage to safely rise by a factor of 10, thus power to rise by a factor of 100.

Another possibility is the fullerene (pure carbon) conductors. For example, 7-12 nm nanotube ropes 200-500 nm in length at ~310 K conduct ~50 nanoamperes at ~1 millivolt, transmitting a total power of ~50 pW. Boron nitride coated carbon nanotube wires should be highly oxidation-resistant. Field emission from room temperature carbon atomic wires has been measured as 0.1-1 microamperes at 80 volts, or ~10 microwatts. Nanowires a few nm wide and hundreds of microns long have already been fabricated, and continuous meter-length threads (and longer) have been proposed; current densities $>10^{14}$ amps/m have been pulled through a single chain of carbon atoms.

If fullerene nanowires are doped with metal atoms, adding electrons or holes so that the charge carrier density is high, electrical conductivity may be precisely engineered from semiconductor levels up to equal or better than copper in the plane of the graphene sheet. Multitube cables may display a quantum confinement effect that sharply reduces the resistance, possibly giving an electrical conductivity 10-100 times that of copper at room temperatures, and may provide shielding comparable to coax. Doped electron-rich conductors are easily oxidized on the nanometer scale, but wrapping such conductors in a second concentric undoped outer nanotube provides an insulating jacket that is effectively chemically impermeable under normal conditions. By 1998, coaxial nanocables ~30 nm in diameter had been fabricated in lengths up to 50 microns. Boron-nitrogen (BN) equivalents of fullerene nanotubes are physically strong and good insulators. Low-temperature superconductivity has been demonstrated in doped and undoped fullerenes.

An interesting alternative is a conveyor belt system for direct mechanical charge transport, crudely analogous to the charging system of a Van de Graff generator. Nanoscale conveyor belts could transport ~10^6 electrons/sec at a belt speed of 4 mm/sec, comparable to the ~1 cm/sec drift speed of electrons flowing in macroscale wires. This gives a 0.2 picoampere current through a device with cross-sectional area ~(10 nm), a modest current density of ~10^3 amperes/m and electrostatic belt stress of ~1 atm between adjacent charges spaced ~4 nm apart. If a much smaller belt can be designed of a scale comparable to the 150 micron long computer data storage tape described by Drexler, a belt speed of ~10 m/sec and charge density of ~1 electron/nm (~amino acid density) on a 1 nm belt gives a current of ~2 nanoamperes and a current density of ~10^9 amperes/m, comparable to the electromigration-limited ~10^{10} amperes/m maximum current density in bulk aluminum, with an electrostatic belt stress of ~3600 atm. Net belt mechanical energy dissipation may be as low as ~10^{-6} zJ/nm of travel for each nanometer-scale charge carrying device.

Electromagnetic Tethers

Optical energy may be piped from one place to another by allowing light to enter a narrow solid fiber of clear plastic or fused silica. Light undergoes total internal reflection at the glass-air boundary and follows the contours of the light pipe, with losses due to scattering and absorption in the material as each photon undergoes up to ~10^5 reflections/ meter during transit. The scattering minimum in commercial ultrapure silica fiberoptic cable occurs at ~1500 nm, or ~2×10^{14} Hz, in the near infrared. These losses are roughly equivalent to transmission in clear air, negligible over distances of relevance in nanomedicine: measurable amounts of energy may be transmitted through single fibers ~100 km long.

At the lowest frequencies, the cutoff frequency for electromagnetic waveguides of diameter d_{guide} filled with material of dielectric constant k_e, below which frequency the waveguide cannot transmit photons (velocity $c = 3 \times 10^8$ m/sec in vacuo, and slightly lower in various dielectric materials at various frequencies), is $n_{cutoff} = c/2k_e\, d_{guide}$ ~ 7.5×10^{13} Hz (1 = 1000 nm) in the near infrared-close to the scattering minimum at 1500 nm—for d_{guide} = 2 microns and k_e ~ 1. This cutoff applies only to waveguides; coax (shielded single wire) or triax (shielded twisted pair) lines can conduct frequencies all the way down to DC and all the way up to near-infrared.

For the highest frequencies, optical fibers may be as small as ~0.51 micron in diameter, roughly the photon wavelength—a dielectric rod can serve as a waveguide crudely analogous to the hollow metal pipes used at rf frequencies. Blue photons (~400 nm) carry ~500 zJ of energy, almost enough to break C-C bonds (~550 zJ). Ultraviolet (UV) wavelengths shorter than ~300 nm are greatly attenuated for fiber lengths >1 meter due to heavy absorption, and intense prolonged UV irradiation of silica creates defects (colour centers) in the material that lead to further absorption of the laser light in the fiber.

In biomedical applications, silica fibers are used to conduct photons from excimer lasers, the brightest known sources of UV radiation. Typical maximum continuous power intensities are ~30,000 watts/m for corneal sculpting, ~10^5 watts/m for bile duct cholangiocarcinoma tumor surgery, and ~10^6 watts/m for arterial debulking, laser dental machining and laser lithotripsy for kidney stones, so ~0.01-1 microwatts may be delivered to medical nanodevices using a single ~1 micron optical tether transmitting UV photons. For a nanomedical optical power cable with A_w = 1 micron diameter fiber, L = 1 meter, T_0 = 310 K and T_{max} < 373 K, then P_{Aw} ~ 10^9 watts/m and a maximum of P ~1 milliwatts may be delivered down the fiber. However, at this high fluence there is tremendous risk of localized tissue incineration, should the fiber detach in vivo and the photon flow is not immediately halted—a good argument for limiting optical power tether intensities to <10^5 watts/m or less. A thin adjacent backchannel fiber could serve as a fuse, rupturing simultaneously with the main fiber and cutting off the feedback control signal.

4

MOLECULES INTO MOTORS

An advanced technology can do away with things we regard as absolutely essential. Just that is happening now in a blossoming field at the intersection of physics, chemistry and biology: the study and construction of devices that serve as motors and pumps on the molecular scale. These mechanisms generally lack rotors, armatures and all the other trappings of conventional engines, but that is the least of their oddities. In an ordinary motor, energy is used to cause motion. In these motors, energy is used to cause a cessation of motion. Although they seem rather like an example of alien technology, they are the most common type of motor on our planet, the basis of the inner workings of all living cells. Our physical intuition, formed by everyday abstention of large machines, fails when we consider the world of the small.

It is a capricious world, ruled by thermal and quantum fluctuations. For molecules, moving deterministically is like trying to walk in a hurricane: the forces propelling a particle along the desired path are puny in comparison to the random forces exerted by the environment. Yet cells thrive. They ferry materials, they pump ions, they build proteins, they move from here to there. They make order out of anarchy. Over the past several years, researchers have finally begun to understand how. The basic insight, loosely described as the Brownian ratchet principle is that random noise can be put to good use. The trick is to reaction the noise, to filter out the randomness you do not want so that you are left with what you do want.

This principle resembles the phenomenon known as *stochastic synchronization* whereby increasing the noise in a communications channel can actually make it easier to transmit a signal.

Using the techniques of chemistry, researchers have been designing miniature motors and devices that can manipulate molecules one at a time. These tiny machines imitate what protein motors and pumps do in living cells—convert chemical energy into mechanical work with almost 100 percent efficiency—and could carry out such tasks as molecular assembly, fine shifting, computation and semi conductor quality control. They may be the first step in turning the science fiction of nanotechnology, the dream of atom-by atom control of matter, into science fact.

Even a freak hailstorm does not come close to the tempestuous bombardment that is routine in the molecular world, but the effects can be analogous. Usually when you park your car at the foot of a bill, turn off the engine and release the emergency brake, the car will not start climbing the hill. But imagine this scenario. Hundreds of *hailstones* strike the car every second, hitting all sides at random. Each one transfers a small amount of momentum to move the car a tiny distance forward or back ward. On average, the momentum transferred to the car is zero, but in any time interval the car will move a little more in one direction than in the other.

You can take advantage of these *random pushes* in a very simple way. Put a brick behind the rear wheel to prevent the car from rolling backward and wait until a hailstone pushes it for ward, If you do nothing, the car will soon roll back, but if you swiftly move the brick, you can trap the car in its new position. By continuing this process—moving the brick each time the car lurches forward—you can drive down the street, even up a bill. It takes a keen eye and quick wit to move a brick under a heaving car in the middle of a violent hailstorm. Fortunately, the same effect can be achieved simply by replacing the Standard brake with a ratchet—a device that allows motion in only one direction. A ratchet consists of a gear with asymmetric teeth and a pawl, a little arm that jams the gear and prevents it from turning backward. In a turnstile or ratchet wrench, the pawl is spring-loaded. Such a simple ratchet mechanism would keep the car moving forward. Progress would be agonizingly slow, however, because only a small fraction of the hailstones—those with enough momentum to overcome the force exerted on the ratchet by the spring—would have any effect.

A better alternative is to skew the teeth in the opposite direction and substitute the spring-loaded pawl with a piston activated by the brake pedal. When the brake is off, the piston is disengaged, and the car is free to lurch back and forth. When the driver pushes the brake pedal, the piston engages and locks the gear (and car) in place. This modified ratchet does away with the need for careful measurement and intelligent intervention. All the driver has to do is sit in the car and pump the brake. Because of the skewed gear teeth, a few extra hailstones striking from behind are sufficient to move the car forward far enough to advance the gear by one tooth, whereas a larger number of hailstones from the front are necessary to push it backward by one tooth. This asymmetry ensures that the car moves even if the brake is engaged and disengaged randomly. The beauty of the system is that it requires no synchronization—none of the careful timing required in an ordinary engine.

Averaged over time, the hailstones exert no net force on the car. The vehicle acquires its forward motion from the application of the brake, which forces the piston down onto the gently sloping face of the lopsided teeth. Take away any component—the asymmetry of the ratchet teeth, the jittering caused by the hailstones or the external energy Supplied by pumping the brake—and the mechanism would fail.

Needless to say, such a contrivance is quite unrealistic for a real car. A hack-of-envelope calculation shows that a reasonable pumping rate could impart a velocity of no more than a kilometer per horn; about a tenth of the cars body length second. The maximum force on the car would he one millionth the gravitational force, so at best the car could climb a very gradual slope. But if the car is very small—say, the size of a large molecule— and immersed in water, the mechanism is much more effective. The mass ratio of a water molecule and a small protein is about the same as the mass ratio of a hailstone and a car. The difference is that water molecules hit the protein many billion times a second. These collisions produce the well-known jittering called *Brownian motion*. What is not so well known is that a Lilliputian ratchet could use Brownian motion to turn directionless energy into directed motion. A small protein could reach a velocity of one micron (more than 10 times its size) a second—the equivalent of 100 kilometers an hour for a car. The ratchet mechanism could overcome a force of up to 10 piconewtons, nearly a million times the force of gravity on a molecule.

It is amazing but true that two random processes can combine to produce a nonrandom effect. Physicist Juan M.R. Parrondo of Complutensian University in Madrid recently showed that the same principle applies to games of chance. Switching between two games, each a losing proposition, can turn the odds in your favour. A physicist's first reaction is that ratchets might break the second law of thermodynamics, whereby it is impossible to convert random thermal fluctuations into mechanical work. In his famous *Lectures on physics*, Richard Feynman analyzed a ratchet attached to a paddle wheel, If the ratchet could prevent the wheel from going backward, molecular collisions would cause an irregular but relentless rotation of the wheel. The result: a perpetual-motion machine of the second kind—that is, one that defies the second law. (The device does not claim to manufacture energy out of nothing, so it does not violate the first law of thermodynamics, the law of conservation of energy.)

As Feynman showed, however, the device cannot work without an outside energy source. The pawl must be attached to the ratchet by a spring, which itself is vulnerable to thermal noise, Occasionally noise causes the spring to contract, lifting the pawl and prematurely disengaging the mechanism, Because of tile asymmetry of the gear teeth, the ratchet will most likely slip back a notch. If the paddle wheel and pawl are at the same temperature, the tendencies to move forward (because of molecular collisions) and to move backward (because of the unreliable spring) exactly cancel. Despite superficial appearances, a ratchet system in thermal equilibrium will not rotate.

This restriction does not apply when the system is out of thermal equilibrium. If the paddle wheel is hotter than the spring, the ratchet rotates counterclockwise, as intuition would suggest. If the spring is hotter, the ratchet rotates clockwise—a motion that ratchets usually prevent. Any departure from equilibrium allows for ratchet-driven motion. Whatever creates the disequilibrium provides energy to the system. In the case of the car, the energy comes from the foot on the brake. The energy dissipates to beat as the car is forced into the locked position. In this way, these systems comply with the second law.

Although large thermal gradients are rare at the molecular level, other forms of disequilibrium are quite common. Three years ago organic chemist T. Boss Kelly and his colleagues at Boston College elaborated on this point with a clever experiment. They synthesized Feynman's ratchet from triptycene, a Y-shaped organic molecule that serves as the paddle wheel, and helicene, a G-shaped molecule that acts as the pawl and spring. Because the helicene has a bend in it, it is easier to turn the paddle wheel clockwise than counterclockwise. Despite this asymmetry, NMB (*Nuclear Magnetic Resonance*) spectroscopy showed that the frequencies of clockwise and counterclockwise turns were exactly equal, as Feynman predicted.

Kelly's group then incorporated a nonequilibrium chemical process: the hydrolysis, or water-driven decomposition, of phosgene gas. A hydroxyalkyl group was attached to the pawl and an amino group to one vane of the paddle wheel. Together they served as a brake. Whenever the vane approached the pawl, the groups (primed by the phosgene) reacted and prevented any further counterclockwise rotation. The net effect was that most of the paddle wheels rotated clockwise. This system is not a true molecular motor—if the brake were released and reapplied, the wheel would tend to return to its starting point—but it does demonstrate the concept. Other groups have achieved continuous rotation using different ratchet mechanisms. A team headed by Ben L Feringa of the University of Groningen, for instance, drives a molecular motor with light.

Recent experiments suggest that at least some *biological engines* work by similar means. One example is the ion pump, a protein that pushes electrically charged particles through a cell membrane. Ions naturally flow from higher to lower electrochemical potential, but these pumps can drive ions in the opposite direction, maintaining the electrochemical gradients essential for life. The ion pump seems to be based on a simpler device, the ion channel. An ion channel is a biological rectifier: it allows electric current to flow in one direction only. A typical channel is a funnel-shaped protein about 10 nanometers long. Ions can move from the mouth of the funnel to the tip, but not the other way. Turning this rectifier into a pump requires some mechanism to modulate the size of the mouth and the strength of the interaction of an ion within the channel. The shape of the channel is well suited to this type of modification, because it acts like a lever: a small displacement of atoms near the tip of the Funnel can result in a large change at the mouth. By making the mouth open and close cyclically, the pump can move ions from the tip of the funnel out the base—just as pumping the car brake caused the gear to turn in the opposite direction from what one would have expected.

The hydrolysis of *adenosine triphosphate* (ATP), the fuel used by cells, provides just the mechanism required to turn a channel into a pump. In a simplified description, the pump has two possible states. In the first state, the mouth is open to the inside of the cell and ions interact strongly with the interior of the channel. In the second, the mouth is closed to the inside and ions interact weakly with the interior. The binding of ATP favours the first state, and the release of the hydrolysis products favours the second. The process is analogous to the operation of a canal lock but with a crucial difference: it requires no control mechanism to synchronize the hydrolysis with the ion motion. It is enough to cycle randomly between two states of the protein. When the gate to the inside is open and the channel's energy level is low, an ion naturally enters the channel from the inside. When the gate to the inside is closed and the energy level is high, the ion flows to the outside. In the mid-1980s this ratchet picture was corroborated by Tian Y. Tsong, then at Johns Hopkins University, me, and our colleagues. We applied an alternating electric field to an ion pump and observed it driving ions up an electrochemical gradient, even without hydrolyzing ATP.

Another example is *kinesin*, a molecular forklift that trans ports proteins within the cell. Kinesin consists of two loosely attached domains and moves along a track called a *microtubule*, made of many individual molecules of the protein tubulin, each about 10 nanometers long. The electric potential between the kinesin and the microtubule usually has a saw tooth pattern, with energy barriers preventing the motion of kinesin from one tubulin molecule to the next. In the Brownian model, hydrolysis of an ATP molecule changes this potential to a flat shape and allows random collisions to jostle the kinesin. Release of the hydrolysis products returns the potential to the sawtooth shape, which, depending on how far the kinesin has drifted, can push the molecule forward.

This Brownian model of how kinesin moves differs radically from the traditional one, in which the shape of the molecule played the central role. The idea was that the two domains, acting like arms, let go of the microtubule one at a time and swing forward—as though they were moving along monkey bars in a playground. A clear prediction of this theory was that if one domain is removed, the resulting molecule should not be able to move along the microtubule. In 1998 Yasushi Okada and Nobutaka Hirokawa of the University of Tokyo replaced one of the domains with a charged loop of amino acids, so that the molecule had a different shape but nearly the same binding energy. This molecule moved roughly as fast as the normal two- armed kinesin, in support of the Brownian model.

Brownian motion may also play a role in the biomolecular motors that make up our muscles, assemble our proteins, synthesize ATP, zip and unzip DNA, transport proteins across our cell membranes, and break down proteins when they are no longer needed. In some cases, the evidence is equivocal, and researchers still debate what is going on. But one thing is clear: any microscopic machine must either work with Brownian motion or fight against it, and the former seems to be the preferable choice.

The Brownian ratchet principle gives scientists and engineers a whole new way to manipulate matter on a small scale. One of the first applications has been to separate particles by weight. This process is a microscopic version of panning for gold. When you subject particles to random fluctuation—either by shaking a tray or by subjecting them to Brownian motion—the heavier ones move at a slower speed. The first steps toward building Brownian sieves were taken in 1992 by physicists Armand Ajdari of the Paris School of Industrial Physics and Chemistry and Jacques Prost of the Curie institute. Recently Joel S. Bader and his colleagues at the biotech company Curagen in New Haven, Conn., built a device to sort DNA molecules. Their approach promises greater precision and selectivity than standard sorting techniques such as electrophoresis, centrifuge and distillation. In all of the above ratchet examples, the ratchet electric field is either on or off.

In 1996, however, Martin Bier suggested using three states: positive, negative or off. By switching among these states, a Brownian sieve could make heavy particles move one way and light particles the other way. Particles could be continuously fed into the middle of the device, collected at either end, fed into another device tuned to a different mass and so on, with ever better separation at each stage. Such devices could sort not just by mass but also by size or electric charge. Theorists at Princeton, Chicago, the Massachusetts Institute of Technology and the University of Ottawa have since extended this idea into two dimensions

Two years ago Alexander van Oudenaarden and Steven G. Boxer, both then at Stanford University, built a working 2-D sieve. They used electron-beam lithography to pattern a glass slide with an array of asymmetric barriers 25 nanometers high. They filled this tiny maze with a fluid of electrically neutral phospholipid molecules, mixed in some phospholipids with various electric charges and applied an electric field. The field pulled the charged molecules through the obstacle course. Because singly charged molecules moved more slowly than doubly charged ones, they had more time to drift sideways while in the space between obstacles. The asymmetric harriers made it easier for them to drift in one direction rather than simply spread out. By the time the charged molecules had reached the other side of the slide, they had sorted themselves into different groups by charge.

It was only a matter of time before ratchets found their way into the quantum world, Pour years ago Peter Hanggi and his colleagues at the University of Augsburg in Germany made the tantalizing suggestion that quantum effects—interference between electron wave functions, quantization of energy levels, electron tunneling through barriers—could provide another randomizing force. These effects would take over from Brownian motion at the lowest temperatures and smallest scales, Using a quantum ratchet, researchers could gain precise control of individual electrons without having to exert comparably precise manipulation of electric fields.

Since that time, Charles M. Marcus, then at Stanford, and his colleagues have made an electron pump from a quantum dot, which acts as a tunnel between two larger reservoirs of electrons and can be closed off by electrostatic gates. By cycling the voltage on the dot and on the gates, Marcus's team pushed electrons between the reservoirs one by one. Because their system was always near equilibrium, the process was reversible, allowing the energy usage to be made arbitrarily small.

Two years ago Imre Derenyi designed a similar mechanism in which the voltage changes would be abrupt and random. Such a system would be intrinsically irreversible—the direction an electron is pumped does not depend on the order in which the steps are carried out—and hence more wasteful. But it would have advantages, especially as a model of irreversible chemical reactions, such as those used to drive the ion pump. Other potential applications include electron pumps in molecular computers and amplification of signals along molecule-scale wires.

Meanwhile Heiner Linke, formerly at Lund University in Sweden, and his colleagues have used triangular quantum dots. The triangles acted as ratchets because it was harder for electrons to squeeze through the vertex. When an oscillating voltage modulated this built-in bias, a net current flowed—even though the average voltage was zero. Varying the temperature regulated the direction of the current. At high temperatures the device functioned as a thermal ratchet: the electrons tended to flow out of the vertex of the triangles because once through the vertex it was harder for them to go hack. At low temperatures it turned into a quantum ratchet: electrons flowed out of the base of the triangles because the width of the energy barrier was smaller in that direction, thereby making tunneling faster. In addition to their applications in electronics, quantum ratchets could be used to damp the current vortices that develop in superconductors, thus resolving a major problem for magnets and superconducting wires. These ideas bring us full circle.

A century ago Brownian motion helped tremendously in demonstrating the existence of atoms. It also explained chemical reaction rates as a balance between thermal noise, which brings molecules together, and electric repulsion, which pushes them apart. The concepts filtered into biology as a possible explanation of biological trans port driven by nonequilibrium chemical reactions. Nowadays biological systems are inspiring the design of chemically synthesized molecular motors and pumps, sophisticated sieves and quantum rectifiers. The flow of ideas has reversed, and the progress is anything but random. In the near future, using the principles of chemistry rather than of mechanical engineering, we may have micron-size factories assembling nanometer-size parts for motors to perform microscopic surgery; pumps to rid the factories (and maybe our cells) of unwanted waste products; and transistors for molecular computers to control these and other processes. Just as in *The War of the Worlds*, which ends with the Martians being felled by humble germs, the small may end up conquering the large.

5

TOP-DOWN MICRO AND NANOTECHNOLOGY

Nanotechnology deals with natural and artificial structures on the nanometer scale, i.e. in the range from 1 μm down to 10 Å. One *nanometer*, 1 nm = 10^{-9} m, is roughly the distance from one end to the other of a line of five neighbouring atoms in an ordinary solid. If the size of a soccer ball ($\sim$30 cm = 3×10^{-1} m) is reduced 10.000 times we reach the width of a thin human hair ($\sim$30 μm = 3×10^{-5} m). If we reduce the size of the hair with the same factor, we reach the width of a carbon nanotube ($\sim$3 nm=3×10^{-9} m).

It is quite remarkable, and very exciting indeed, that we today have a technology that involves manipulation of the ultimate building blocks of ordinary matter: single atoms and molecules.

Nanotechnology owes it existence to the astonishing development within the field of micro electronics. Since the invention of the integrated circuit nearly half a century ago in 1958, there has been an exponential growth in the number of transistors per micro chip and an associated decrease in the smallest width of the wires in the electronic circuits. As a result extremely powerful computers and efficient communication systems have emerged with a subsequent profound change in the daily lives of all of us.

A modern *computer chip* contains more than 10 million transistors, and the smallest wire width are incredibly small, now entering the sub 100 nm range. Just as the American microprocessor manufacturer, Intel, at the end of 2003 shipped its .first high-volume 90 nm line width production to the market, the company announced that it expects to ramp its new 65 nm process in 2005 in the production of static *RAM* chips. Nanotechnology with active components is now part of ordinary consumer products.

Conventional microtechnology is a top-down technology. This means that the microstructures are fabricated by manipulating a large piece of material, typically a silicon crystal, using processes like *lithography*, *etching*, and *metallization*. However, such an approach is not the only possibility. There is another remarkable consequence of the development of micro and nanotechnology.

Since the mid-1980s a number of very advanced instruments for observation and manipulation of individual atoms and molecules have been invented. Most notable are the *atomic force microscope* (AFM) and the *scanning tunnel microscope* (STM) that will be treated later in the lecture notes. These instruments have had en enormous impact on fundamental science as the key elements in numerous discoveries. The instruments have also boosted a new approach to technology denoted bottom-up, where instead of making small structure out over large structures, the small structures are made directly by assembling of molecules and atoms.

Microfabrication and Moore's Law

The top-down approach to microelectronics seems to be governed by an exponential time dependence. In 1965, when the most advanced integrated circuit contained only 64 transistors, Gordon E. Moore, Director of Fairchild Semiconductor Division, was the first to note this exponential behaviour in his famous paper *Cramming more components onto integrated circuits*. "When unit cost is falling as the number of components per circuit rises, by 1975 economics may dictate squeezing as many as 65,000 components on a single silicon chip". He observed a doubling of the number of transistors per circuit every year, a law that has become known as *Moore's law*.

Today there exist many other versions of Moore's law. It concerns the exponential decrease in the length of the gate electrode in standard CMOS transistors, and relates to the previous quoted values of 90 nm in 2003 and 65 nm in 2005. Naturally, there will be physical limitations to the exponential behaviour expressed in Moore's law. However, also economic barriers play a major if not the decisive role in ending *Moore's law* developments. The price for constructing microprocessor fabrication units also rises exponentially for each generation of microchips. Soon the level is comparable to the gross national product

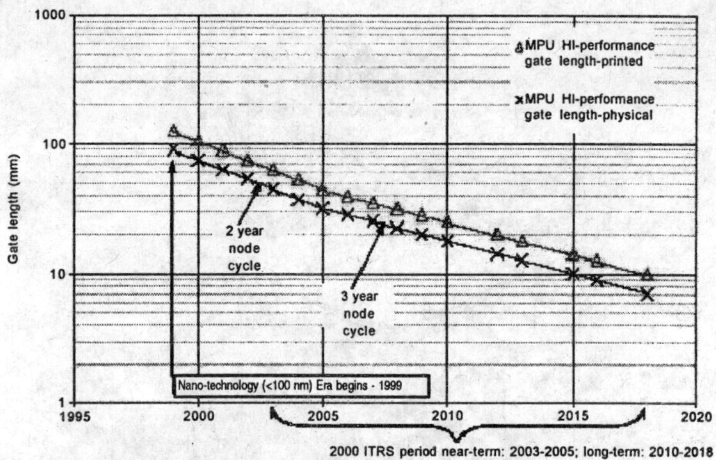

Fig. 5.1. Moore's law applied to the shrinking of the length of the gate electrode in CMOS transistors.

of a mid-size country, and that might very well slow down the rate of progress.

Clean Room Facilities

The small geometrical features on a microchip necessitates the use of clean room facilities during the critical fabrication steps. Each cubic meter of air in ordinary laboratories may contain more than 10^7 particles with diameters larger than 500 nm. To avoid a huge flux of these "*large*" particles down on the chips containing micro and nanostructures, micro and nanofabrication laboratories are placed in so-called clean rooms equipped with *high-efficiency particulate air* (*HEPA*) filtering system. Such systems can retain nearly all particles with diameters down to 300 nm. Clean rooms are classified according to the maximum number of particles per cubic foot larger than 500 nm. Usually a class-1000 or class-100 clean room is sufficient for *microfabrication*.

The low particle concentration is ensured by keeping the air pressure inside the clean room slightly higher than the surroundings, and by combining the *HEPA* filter system with a laminar air flow system in the critical areas of the clean room. The latter system let the clean air enter from the perforated ceiling in a laminar flow and leave through the perforated floor. Moreover, all personnel in the clean room must be wearing a special suit covering the whole body to minimize the surprisingly huge emission of small particles from each person.

Fig. 5.2. The clean room facilities DANCHIP, situated next to MIC.

The air flow inside the *DANCHIP* clean room is about 1.3×10^5 $m^3 h^{-1}$, most of which is recirculated particle-free air from the clean room itself. However, since the exhaust air from equipment and fume hoods is not recirculated, there is in intake of fresh air of 0.3×10^5 m^3 h^{-1}.

Photolithography

Almost all top-down manufacturing involves one or more *photolithography fabrication* steps, so we give a brief outline of this technique here. A generic *photolithography* process is sketched in Fig. 5.3. From a lightsource light is directed through a mask carrying the circuit design down onto the substrate wafer covered with a photosensitive film, denoted the photoresist. Depending on the local photo-exposure defined by the photolitographic mask the photoresist can be partly removed by a chemical developer leaving well-defined parts of the substrate wafer exposed to etching or metal deposition.

The *substrate wafer* is typically a very pure silicon disk with a thickness around 500 μm and diameter of 100 mm (for historic reasons denoted a 4 inch wafer). Wafers of different purities are purchased at various manufacturers.

The photolithographic mask contains (part of) the design of the microsystem that is to be fabricated. This design is created using *computer-aided design* (*CAD*) software. Once completed the computer file containing the design is sent to a company producing the mask. At

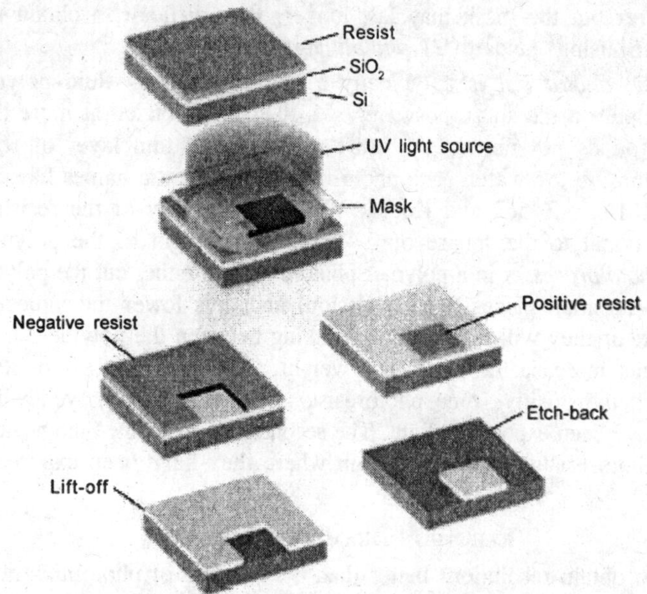

Fig. 5.3. The basic principles of photolithography.

the company the design is transferred to a glass plate covered with a thin but non-transparent layer of chromium. The transfer process is normally based on either the relatively cheap and fast laser writing with a resolution of approximately 1.5 μm and a delivery time of around two weeks, or the expensive and rather slow electron beam writing with a resolution of 0.2 μm and a delivery time of several months.

The photo exposure is typically performed using the 356 nm UV line from a mercury lamp, but to achieve the line widths of sub 100 nm an extreme UV source or even an X-ray source is needed. To achieve the best resolution must minimize note only the wavelength λ of the exposure light, but also the distance d between the photolithographic mask and the photoresist-covered substrate wafer, and the thickness t of the photoresist layer. The minimum line width *wmin* is given by the approximate expression

$$w_{min} = \frac{2}{3}\sqrt{\lambda(d+t)}. \qquad \text{...(1)}$$

If $d = 0$ nm the mask is touching the photoresist. This situation, denoted contact printing, improves the resolution but wears down the mask. If $d > 0$ nm, a case denoted proximity printing, the resolution is pourer but the mask may last longer. It is difficult to obtain w_{min} < 2 μm using standard *UV photolithography*.

The *photoresist* is a typically a melted and thus fluid polymer that is put on the substrate wafer, which then is rotated at more than 1000 rounds per minute to ensure an even and thin layer of resist spreading on the wafer. The photoresists carry exotic names like SU-8, PMMA, AZ4562 and Kodak 747. The solubility of the resists is proportional to the square of the molecular weight of the polymer. The *photo-processes* in a polymer photoresist will either cut the polymer chains in small pieces (chain scission) and thus lower the molecular weight, or they will induces cross-linking between the polymer chains and thus increase the molecular weight. The first type of resists is denoted the positive tone photoresists, they will be removed where they have been exposed to light. The second type is denoted the negative tone photoresists, they will remain where they have been exposed to light.

Electron Beam Lithography

To obtain resolutions better than the few μm of photolithography it is necessary to use either *X-ray lithography* or *electron beam lithography*. Here we give a brief overview of the latter technique.

After development of the resist one can choose to etch the exposed part of the wafer. Acid will typically not etch the polymer photoresist but only the substrate, so etching will carve out the design defined by the mask. The shape of the etching depends on the acid and the substrate. It can be isotropic and have the same etch rate in all spatial directions, or it can by anisotropic with a very large etch rate in some specific directions. One can choose the etching process that is most suitable for the design.

Metal deposition followed by lift-off is another core technique. Here a thin layer of metal (less than 500 nm) is deposited by evaporation technique on the substrate after developing the resist. At the exposed places the metal is deposited directly on the substrate, and elsewhere the metal is residing on top of the remaining photoresist. After the metal deposition the substrate is rinsed in a chemical that dissolves the photoresist and thereby lift-off the metal residing on it. As a result a thin layer of metal is left on the surface of the wafer in the pattern defined by the photography mask.

The above mentioned process steps can be repeated many times with different masks and very complicated devices may be fabricated that way.

Electron beam lithography is based on a *electron beam microscope*, in which a focused beam of fast electrons are directed towards a resist-covered substrate. No mask is involved since the position of the electron beam can be controlled directly from a computer through electromagnetic lenses and deflectors.

The electrons are produced with an electron gun, either by *thermal emission* from hot *tungsten filament* or by *cold field emission*. The emitted electrons are then accelerated by electrodes with a potential $U \approx 10$ kV and the beam is focused by *magnetic lenses* and steered by *electromagnetic deflectors*.

The electron is both a particle and a wave. The wavelength λ of an electron is given in terms by the momentum p of the electron and *Planck's constant h* by the *de Broglie relation* $\lambda = h/p$. In the electron beam microscope the electron acquires a kinetic energy given by the acceleration voltage U as $\frac{1}{2}mv^2 = eU$, where m and $-e$ is the mass and charge of the electron, respectively. Since $p = mv$ the expression for the wavelength λ becomes

$$\lambda = \frac{h}{\sqrt{2meU}}, \qquad \text{...(2)}$$

which for a standard potential of 10 kV yields $\lambda = 0.012$ nm.

However, the resolution of an electron beam microscope is not given by λ. First of all, one can not focus the electron beam on such a small length scale. A typical beam spot size is around 0.1 nm. But more importantly are the scattering processes of the electrons inside the resist and the substrate. As illustrated by the computer simulation, the backscattering of the electrons implies that an area much broader area is exposed to electrons than the area of the incoming electrons. This results in an increase of the resolution. It turns out that in practice it is difficult to get below a minimum linewidth of 10 nm.

Electron beam lithography is still the technique with the best resolution for lithography. A major drawback of the method is the long expose time required to cover an entire wafer with patterns. The exposure time *t*exp is inversely proportional to the current I in the electron beam and proportional to the clearing dose D (required charge per area) and the exposed area A,

$$t_{exp} = \frac{DA}{I}. \qquad \text{...(3)}$$

In photolithography the entire wafer is exposed in one flash, like parallel processing, whereas in electron beam lithography it is necessary to write one pattern after the other in serial processing. For mass production electron beam lithography is therefore mainly used to fabricate masks for photolithography.

Nanoimprint Lithography

Nanoimprint lithography (*NIL*) is a relatively young technique compared to photolithography and electron beam lithography. The first results based on NIL was published in 1995.

The technique is in principle very simple. It consists of pressing machine that presses a stamp or mold containing the desired design down into a thin polymer film spun on top of a wafer and heated above its glass transition temperature.

Naturally, a stamp is needed, and often its is produced by making a nanostructured surface in some wafer by use of electron beam lithography and subsequent etching techniques. Different materials have successfully been used as stamps among them silicon, silicon dioxide, metals, and polymers. Often it is necessary to coat the stamp with some anti-stiction coating to be able to release the stamp form the target material after pressing.

The target material is a polymer, which is useful for two reasons. First, above the glass transition temperature polymers are soft enough to make imprinting possible. Second, polymers can be functionalized to become sensitive various electric, magnetic, thermal, optical and biochemical input. Thus the resulting nanostructure can become very sophisticated indeed.

The *pressing machine* needs be able to deliver the necessary pressure. Moreover, it must contain an efficient temperature control in the form of heater plates and a thermostat, since it is crucial to operate at the correct temperature somewhat above the glass transition

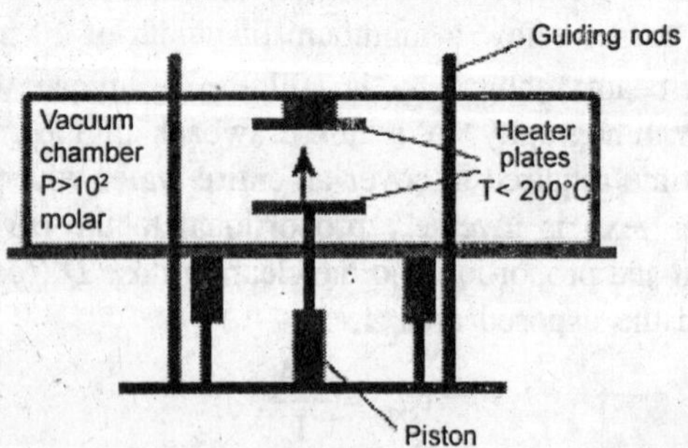

Fig. 5.4. A sketch of the pressing machine at MIC to used to perform nanoimprint lithography.

temperature of the polymer. To avoid impurities it must also operate under a sufficiently low vacuum, and finally it should allow for correct alignment of the sample before pressing.

Nanoimprint lithography is one of the few nanotechnologies that seems to be capable of mass production. Once the stamp is delivered, and the pressing machine is correctly set up, it should be possible to mass fabricate nanostructured wafer. The cycle time of a typical nanoimprint machine is of the order of minutes. This time scale is determined by the actual time it takes to press the stamp down and the various thermal time scales for heating and cooling of the sample.

6

Analyzing Nanomachines

The most basic theory relevant to chemistry is called *quantum electrodynamics*, or *QED*. In simple terms, QED tells us how photons and electrically charged particles, like electrons or atomic nuclei, behave. Physicists, of course, are interested in deeper stuff like how the nucleus behaves internally, but we can get along just fine without that. In fact, we can get along fine without a lot of QED itself. This is good, because calculations using it are quite difficult. It would be a bit like using the theory of relativity to calculate where a baseball is going. There is no practical difference between the results of the much simpler calculation using *Isaac Newton's laws* of motion and the more difficult relativistic one. In the same way, for purposes of chemical interactions, we can almost always get away with using the *Schrodinger equation*. This still captures most of the quantum effects.

The quantum world is strange and hard to understand. Our instincts about the way things work are formed in the larger world, where objects exist, and waves have to be waves in something, like sound in air, or at the surface of water. In the quantum world, both of these descriptions apply to some degree to everything. Something can seem to be an object, like an electron, but can act as if it were in many places, or moving in many different directions, or had many different energies at once. This odd property is called superposition of states and is the thing that will let quantum computers be able to calculate many different things at the same time, thus being vastly more powerful than ordinary computers.

By the same token, when you take an ordinary computer and try to simulate something in the quantum world, the calculations can easily get out of hand. You have to keep track of all the superposed states and the interactions between all the particles. The amount of work blows up exponentially with the number of particles. Therefore, even when doing the quantum mechanics of a chemical interaction, the analysis is limited to a relatively small number of atoms, and something called the *Born-Oppenheimer approximation* is used. This approximation treats the nuclei as if they were ordinary objects and calculates the electrons quantum mechanically around them. Nuclei, it turns out, are heavy enough to act like ordinary objects most of the time.

Even so, a study such as that done by Robert Freitas and Ralph Merkle of the reactions involved in depositing carbon atoms to build up diamond machine parts required 50,000 CPU-hours of calculation.

We know that the Born-Oppenheimer approximation doesn't always hold. Physicists can do experiments that cause superposition with entire atoms being in two places at once, or meld together in overlapping states called *Bose-Einstein condensates*. But this is where the distinction between science and engineering is important. A can design machines that don't do those experiments, and only do the things we can analyze (relatively) straightforwardly instead. We know that atoms act like objects under most normal conditions—we can image them with scanning probes, push them around, put them somewhere, and find them still there when we look back.

Molecular Mechanics

If you are an automotive engineer, you need to analyze the chemistry in the fuel-air explosions in the cylinders of your motor, but you don't need to simulate the chemistry inside your structural steel to know that it remains steel from moment to moment. You use a higher-level approximation that involves weight, tensile strength, coefficients of elasticity, and so forth. The higher-level approximation that is appropriate for the parts of nanomachines where chemistry is not being done is called *molecular mechanics*.

Molecular mechanics treats atoms as objects and condenses the descriptions of what the electrons do around and between their into a number of simpler forces. First is the *van der Waals force*, which is a small attraction between atoms up to a point, and a larger force resisting their being pushed together after that. The van der Waals force makes atoms act like objects, somewhat soft and springy, but with a definite size.

Next are the *ionic forces*, which occur when atoms gain or lose electrons (or part of one, in a molecule). The resulting electric charge

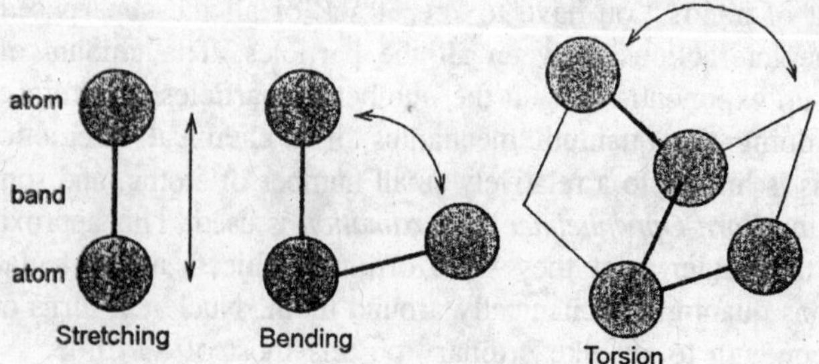

Fig. 6.1. Molecular mechanics. The forces along and between covalent bonds that are accounted for in a typical molecular mechanics force field.

causes a longer-range attraction or repulsion between atoms. For example, ordinary table salt is a crystal of sodium and chlorine ions in which the sodium have lost electrons and the chlorines have gained them. (*Ion* simply means an atom that has lost or gained one or more electrons.) The electrostatic forces that result make the atoms form a very strictly alternating pattern in the crystal. In the *water molecule*, H_2O the oxygen atom pulls the electron from each hydrogen in dose, so that the hydrogen's nucleus (a single proton) is left sticking out a little. The result is that the hydrogen act positively charged and the oxygen acts negatively charged, though not as strongly as if the electron had been pulled all the way off. Thus water molecules stick to each other, hydrogen to oxygen, and stick to the ions of salt, which is why salt dissolves so well in water.

Finally comes the force associated with a *covalent bond*. This is where, instead of one atom grabbing the electron and the other letting go, they both hold on more or less equally. (Actually, the electrons are shared in pairs.) A covalent bond acts as if there were a strong spring between the atoms, holding them much closer together than the van der Waals force would let them, because they actually overlap where the shared electrons are.

Covalent bonds are much stronger than ionic forces, and ionic forces in turn are much stronger than van der Waals ones. Diamond, for instance, is a crystal held together with covalent bonds, and salt is one held together ionically. For a van der Waals crystal, you'd have to look at solid xenon, which melts at —169° Fahrenheit. Stronger solids are held together by van der Waals forces between molecules, but they get help by having the molecules themselves held together by covalent bonds. Ordinary ice is a crystal held together by the weaker ionic bonds between the hydrogen of one molecule and the oxygen of others (which are often called *hydrogen bonds*).

Molecular mechanics describes what goes on in and between molecules at this level of detail. A particular set of parameters describing the strengths and behaviours of the forces and bonds in molecular mechanics is commonly called a force field or sometimes a potential, short for "*potential energy*." (The potential energy of two atoms with a force between them varies with their distance, just as the potential energy of a heavy object increases with its height above the ground.) Potential and force are two sides of the same coin; force is just the derivative (steepness of slope) of potential. In studying the same molecule, a chemist is likely to be more interested in the potential and a mechanical engineer more concerned with the force.

Van der Waals and ionic forces are pair wisc and radial; that is, they cause forces between two atoms to push them directly toward or away from each other. Covalent bonds are more complex. They have a radial, pairwise component that is typically called "*stretching*."

Then a component called "*bending*" depends on the angle between two bonds on a given atom. A component called "*torsion*" depends on the angle between the planes formed by three bonds (connecting four atom). The actual magnitude of the force depends on which elements ire involved, and sometimes on what other elements they are bonded to. For example, the length of a carbon-carbon bond depends on whether the carbon is bonded to three other carbons, as in graphite, or four carbons, as in diamond. Various formulations of force fields have other bells and whistles, but these are the main features.

Most force fields were developed by chemists for different reasons, using different methods to arrive at their parameters. Many are line-tuned to work best in the aqueous environment of a cell's interior and are used for analyzing DNA, proteins, drugs, and so on. Others are optimized for the study of crystals, metals, plastics, and so forth. The reason tuning matters is that the force field is, after all, an approximation to the true quantum mechanical situation, and in a different situation a different approximation may be a little closer.

To a chemist, "a little closer" makes a big difference. For example, a very small difference in the potential of a protein folded into shape A as compared with shape B can mean that in the cell or test tube, virtually all the protein molecules will be in shape A and almost none in shape B. Protein molecules are floppy, with many places where just one bond joins one part of the molecule and the next. The amount of energy necessary to cause that bond to turn one way rather than another is very small. The nanomechanical engineer, however, can use the difference between science and engineering to good effect. We can design stiff molecules where there is at least a tripod of bonds holding each atom in place. The amount of energy necessary to change the shape of such a molecule is much greater. This means that if there is a small inaccuracy in the force field, it hardly changes the shape of the molecule at all, where it would make the protein completely different.

For example, the molecular mechanical engineer can use, for preliminary work, a force field that only describes bond stretching and bending. *Bond torsion* is a correction of about 1%, and thus can be ignored for a first approximation. Any molecule where bond torsion would make a significant difference is nowhere near stiff enough for use as a molecular machine part. In contrast, solution chemists often use an approximation where the bond stretch is not modeled, but an average length used as if it were completely rigid. They can do this because in solution, there are no forces strong enough to change the average length significantly.

Imagine simulating a steel chain. If you are a chemist, you are trying to predict what shape of a heap it will make when you drop it on the floor. You need to be very exact about how much friction there is between links, how they bounce off one another as they clatter down, and so forth. You can assume each steel link will remain the same size and shape, though, and worry only about the smaller forces that move them around. If you are a molecular engineer, however, you are simulating the chain being used to lift a ten-ton weight. You can bet that the links will hang in a straight line from the winch to the weight; but you had better be concerned about how they stretch and deform as the weight nears their holding capacity.

The molecular mechanics force held, together with the description of the particular molecule you are interested in, lets you calculate a lot of things about it. It will tell you what shape it will take, and how hard it will be to bend, and what its vibrational frequencies will be. The other thing you can do is simulate it in what is called molecular dynamics.

In molecular dynamics, you take a molecule or a set of them, and give each atom a velocity, chosen at random, to simulate the kind of velocities they'd have due to heat at, say, room temperature. Then you use the force held to calculate the force on each one, and figure where it would have moved to a femtosecond or some traction thereof later. Now make a movie, doing this entire calculation once per frame.

Watching this movie lets you see if the machine you designed does what you thought it would. Molecular dynamics is peculiarly suited to simulating nanomachines as contrasted with many of the other things, like protein activity, that it is used for. The reason is that things happen much faster in a nanomachine, and the glimpse you get with a simulation is brief.

A femtosecond is a unimaginably short amount of time. If you made a movie in which each successive frame represented a femtosecond of simulated time, it would take a million years to watch one second of simulated time. The speed of sound in diamond is faster than ten miles per second, though, so it would only take two seconds of viewing time for a vibration to move across a one nanometer part.

Most existing molecular dynamics software, and the force fields it implements, is aimed at chemists and materials scientists. That is, it typically treats the group of atoms it is simulating as a sample of a much larger amount of material, and it is used to determine things like heat capacity, pressure-volume-temperature relationships, and so forth. For simulating nanomachines, however, we want to treat molecules as machine parts and measure such things as stiffness, torque, and power. New molecular mechanics and dynamics software is being developed specifically for the purpose of simulating and analyzing mechanical nanomachines.

NANODESIGN

Nanomachines are going to be among the most complex things that humans have ever designed. Grains of beach sand range in size from about 100 microns to about a millimeter. A nanomachine the size of a small grain of sand might easily contain five trillion parts, many of which would be moving parts. If you spread the parts out on a table to get a look at them before trying to put the machine together, and you magnified them only to the size of grains of sand, your table would have to be bigger than 500 square feet. If you magnified them to the size of typical machine parts in a car, the table would have to be bigger than 100 acres.

VLSI (very large scale integration) chips such as microprocessors are among today's most complex designed machines, with millions of working parts. Even these are much too complex to design by hand, even for a large, well-organized engineering team. Instead the team relies heavily on design and simulation software. There are a number of strong parallels between microchips and nanomachines. First, you can simulate them at many different levels of detail and accuracy, with more accuracy being more computationally expensive. On a chip, you can simulate the transistors and wires as three-dimensional pieces of material and simulate the electron density flowing through them. Once you have transistors that work the way you want, you tend to have simpler models for them and the wires. You can simulate the generation of electromagnetic waves by the flowing current or ignore it. You can treat the transistors as simple on/off switches, and either simulate or not the delay times for signals to move through them. You can simulate just the logic, where small groups of transistors are represented as Boolean functions such as AND and OR, the primitive one-bit operations of which the computer logic is built at this ultra simplified level, a microprocessor is fiendishly complicated.

Each of these levels of simulation will properly represent the machine if the science/engineering distinction is observed. In other words, each level of simulation makes some assumptions about the design at the levels below it and only gives valid results if those assumptions are true. It is up to the designer to make sure that they are.

In electronics design, the assumptions are ensured in a couple of ways. First, designers often make use of predesigned parts that have been carefully analyzed at the lower levels. For example, someone doing logic design will use logic gates that have been carefully checked out as electronic circuits to have guaranteed limits on signal rise and fall times, propagation delay, and so forth. Then the logic designer need only be concerned with the Boolean logic and not have to worry about electronic engineering as well.

Second, the CA software that designers use contains subsystems called *design-rules checkers*. These are very similar to the spelling checkers that are of text—editing programs. If you are drawing a circuit as a direct on your screen, a design rule may say you cannot put two wires closer than some given distance apart. You do not need to know the low—level analysis that showed inductive coupling if they were too close, or that the diffraction limit of the fabrication process just won't let you actually build them closer. All you need to know is that when you draw them too close, they get outlined in red boxes and you must move them apart.

With libraries of prepackaged modular parts and automatic design rules checking, you can be confident that circuits designed at quite an abstract level will work as intended. Exactly the same techniques will be used to design nanomachines.

Today, when people design physical mechanical machines, they pay a lot of attention to each part, optimizing function, manufacturability, and cost. This is very much like how integrated circuits were designed in the 1960s and 1970s. There was enormous attention to detail, and lots of analysis of the effects of each part on every other part. Then circuits began getting more and more complex, as integrated circuits (ICs) became large scale integration (LSI), which became very large scale integration (VLSI). The complexity of a chip rapidly hit and surpassed the limits of what a human brain could understand. A semiserious joke in the industry in those clays was that any electronics engineer had two chips in him—after he'd designed two, he was burned out and fit only for management.

There are three broad ranges in the complexity of machines. The first is like a person living alone, with a bedroom, wardrobe, kitchen, garage, and car. Everything simple and straightforward: when you need the can you drive it; when you need to cook, you go to the kitchen. The second stage is more like having a large family living in the same circumstances. Now everything is being used more efficiently, at the cost of keeping track of trade-offs and interactions. Someone is always in the kitchen, and junior is doing chemistry experiments in the sink between meals. Dad gets driven to work because Mom needs the car during the day. You almost need a reservation to use the bathroom.

The third stage is like a large company. Each person gets facilities that are appropriate to his task. Sharing and interaction are planned and are nowhere as helter-skelter as in the family. There are what seem to be large inefficiencies: the office building sits unused all night, the cafeteria is unused much of the day, where with the family, homework is done in bedrooms and on the kitchen table. Still, every thing works well enough.

In electronics, the first stage is the *discrete transistor* in its own package. In mechanical engineering, it is the craftsman building a single tool. The second stage is the *IC* of the 1970s or the mechanical system of today. Lots of ingenuity goes into getting everything into the constrained resources and making it work together. The giant VLSI chips of today, and the astronomically complex nanomechanical systems of tomorrow, represent the third stage. In the third stage, the bottleneck is not the physical resources of the system, but the ability to design and direct. In the company, it's often cheaper to get extra rooms, machines, whatever, than to buy fewer ones but hire another manager to decide how they to be shared efficiently. In *VLSI*, modules are pre designed and used as is, even though each one might have been further optimized for its specific function. Chip layouts are regularized and laid down by automatic programs, even though carefully (and time-consuming) human attention might have clone it a little better. Indeed, chip designs today look more like programs, written in languages like VHDL and Verilog, than they do any graphical picture of the chip.

Stage three for nanomachines will follow the same route. A mechanical engineer of today; looking at (any small part of an advanced nanomachine, would consider its design quite sketchy and pedestrian. The reason is that it will be generated automatically as part of a vast system, probably described in something like a programming language instead of anything three-dimensional. The designer will have written, essentially, what she wanted the machine to do, rather than trying to describe the low-level mechanism to do it.

Indeed, the designer might not even know (or care) whether she is describing hardware or writing software. Suppose she is designing a machine that will perform a certain motion. She describes the motion she wants, and the design software decides whether it will be a clever linkage, a more straightforward linkage driven by cams, a general arm driven by stepper motors controlled by a microprocessor, or some combination.

Several kinds of current-day software have capabilities that promise the kind of design capabilities that we will need for big nanomechanical systems. There are the VLSI design systems.

Another, more common example is software development systems. Back in the 1970s, Fortran and Cobol were the tools many developers used in scientific and business programming, respectively, but a preponderance of work was still done in assembly language, specifying low-level machine instructions one at a time. Today, a vast array of software synthesis techniques is available. Specialized programming languages for almost any imaginable purpose are available, which cut the effort to develop a program by orders of magnitude compared to assembly language. For highly stylized applications, systems that are essentially automated questionnaires let you describe the application you need, and generate it automatically. And there are methods at virtually every point between these extremes.

Simulation and Testing

Today, the best we can do to test a nanomechanical design is simulate it. In the future, however, it will he possible to build and test systems of considerable complexity, like the trillion-parts grain of sand, in seconds. Design practice has tended to move from physical testing to simulation over the past several decades, as computers have become cheaper and faster. In the next few decades, however, the ability to make things physically will follow the same track. For certain applications anyway, there might be a swing back in the opposite direction.

Imagine a complete machine shop, factory, and engineering laboratory in a building with ten acres of floor space. It has every conceivable kind of construction and test equipment, overhead cranes, mobile robots—the works. Reduced to the scale of nanomachines, this fits in a 40-micron square. It would sit comfortably on the cut end of the finest human hair. Your nanoengineering design workstation could have thousands of these facilities built right into the processor. Having specified your design, the workstation could build and test thousands of minor variants in the time it took to press the return key.

7

REPLICATORS

On March 27, 1981, CBS radio news quoted a NASA scientist as saying that engineers will be able to build self-replicating robots within twenty years, for use in space or on Earth. These machines would build copies of themselves, and the copies would be directed to make useful products. He had no doubt of their possibility, only of when they will be built. He was quite right. Since 1951, when John von Neumann outlined the *principles of self-replicating machines*, scientists have generally acknowledged their possibility. In 1953 Watson and Crick described the structure of DNA, which showed how living things pass on the instructions that guide their construction. Biologists have since learned in increasing detail how the self-replicating molecular machinery of the cell works. They find that it follows the principles von Neumann had outlined. As birds prove the possibility of flight, so life in general proves the possibility of self-replication, at least by systems of molecular machines. The NASA scientist, however, had something else in mind.

CLANKING REPLICATORS

Biological replicators, such as viruses, bacteria, plants, and people, use molecular machines. *Artificial replicators* can use bulk technology instead. Since we have bulk technology today, engineers may use it to build replicators before molecular technology arrives. The ancient myth of a magical life-force (coupled with the misconception that the increase of entropy means that everything in the universe must constantly run down) has spawned a meme saying that replicators must violate some natural law. This simply is not so. Biochemists understand how cells replicate and they find no magic in them. Instead, they find machines

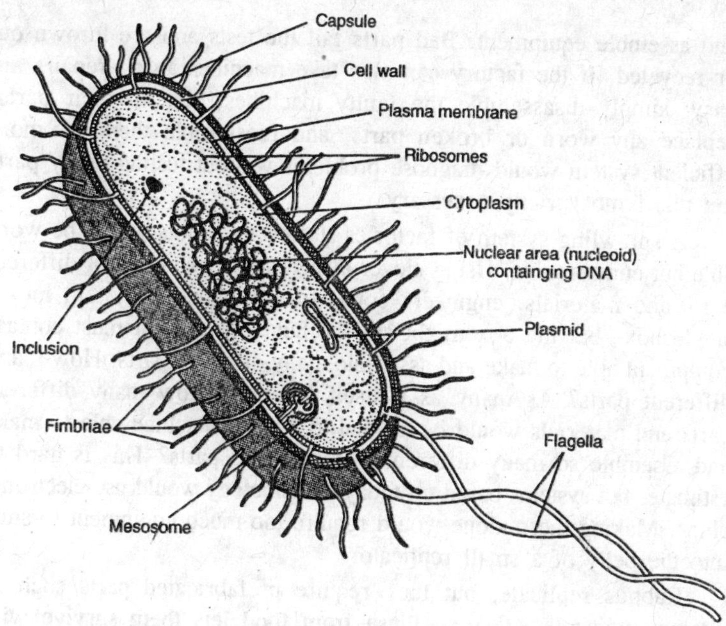

Fig. 7.1. Bacteria.

supplied with all the materials, energy, and instructions needed to do the job.

Cells do replicate; robots could replicate. *Advances in automation* will lead naturally toward mechanical replicators, whether or not anyone makes them a specific goal. As competitive pressures force increased automation, the need for human labour in factories will shrink. Fujitsu Fanuc already runs the machining section in a manufacturing plant twenty-four hours a day with only nineteen workers on the floor during the day shift, and none on the floor during the night shift. This factory produces 250 machines a month, of which 100 are robots. Eventually, robots could do all the robot-assembly work, assemble other equipment, make the needed parts, run the mines and generators that supply the various factories with materials and power, and so forth. Though such a network of factories spread across the landscape wouldn't resemble a pregnant robot, it would form a self-expanding, self-replicating system. The assembler breakthrough will surely arrive before the complete automation of industry, yet modem moves in this direction are moves toward a sort of gigantic, *clanking replicator*.

But how can such a system be maintained and repaired without human labour? Imagine an automatic factory able to both test parts and assemble equipment. Bad parts fail the tests and are thrown out or recycled. If the factory can also take machines apart, repairs are easy: simply disassemble the faulty machines, test all their parts, replace any worn or broken parts, and reassemble them. A more efficient system would diagnose problems without testing every part, but this is not strictly necessary.

A sprawling system of factories staffed by robots would be work able but cumbersome. Using clever design and a minimum of different parts and materials, engineers could fit a replicating system into a single box—but the box might still be huge, because it must contain equipment able to make and assemble many different parts. How many different parts? As many as it itself contains. How many different parts and materials would be needed to build a machine able to make and assemble so many different materials and parts? This is hard to estimate, but systems based on today's technology would use electronic chips. Making these alone would require too much equipment to stuff into the belly of a small replicator.

Rabbits replicate, but they require prefabricated parts such as vitamin molecules. Getting these from food lets them survive with less molecular machinery than they would need to make everything from scratch. Similarly, a mechanical replicator using prefabricated chips could be made somewhat simpler than one that made everything it needed. Its peculiar "*dietary*" requirements would also tie it to a wider "*ecology*" of machines, helping to keep it on a firm leash. Engineers in NASA-sponsored studies have proposed using such semi-replicators in space, allowing space industry to expand with only a small input of sophisticated parts from Earth.

Still, since bulk-technology replicators must make and assemble their parts, they must contain both part-making and part-assembling machines. This highlights an advantage of molecular replicators: their parts are atoms, and atoms come ready-made.

Molecular Replicators

Cells replicate. Their machines copy their DNA, which directs their ribosomal machinery to build other machines from simpler molecules. These machines and molecules are held in a fluid-filled bag. Its membrane lets in fuel molecules and parts for more nano machines, DNA, membrane, and so forth; it lets out spent fuel and scrapped components. A cell replicates by copying the parts inside its membrane bag, sorting them into two clumps, and then pinching the bag in two. *Artificial replicators* could be built to work in a similar way, but using assemblers instead of ribosomes. In this way, we could build cell-like replicators that are not limited to molecular machinery made from the soft, moist folds of protein molecules.

But engineers seem more likely to develop other approaches to replication. Evolution had no easy way to alter the fundamental pat tern of the cell, and this pattern has shortcomings. In synapses, for example, the cells of the brain signal their neighbours by emptying bladders of chemical molecules. The molecules then jostle around until they bind to sensor molecules on the neighbouring cell, some times triggering a neural impulse. A chemical synapse makes a slow switch, and neural impulses move slower than sound. With assemblers, molecular engineers will build entire computers smaller than a synapse and a million-fold faster.

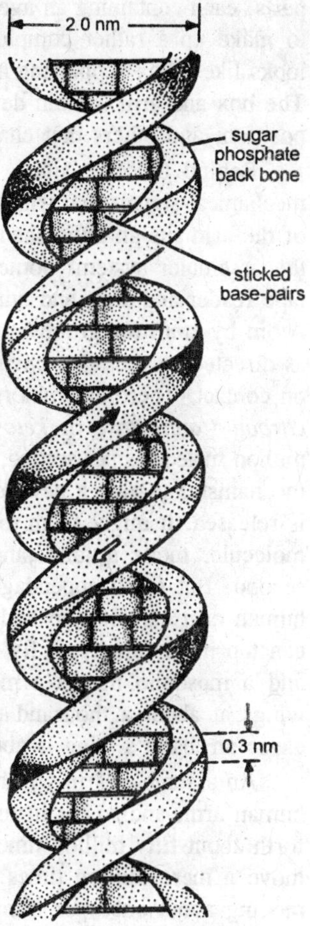

Fig. 7.2. DNA.

Mutation and selection could no more make a synapse into a *mechanical nanocomputer* than a breeder could make a horse into a car. Nonetheless, engineers have built cars, and will also learn to build computers faster than brains, and replicators more capable than existing cells. Some of these replicators will not resemble cells at all, but will instead resemble factories shrunk to cellular size. They will contain nanomachines mounted on a molecular framework and conveyor belts to move parts from machine to machine. Outside, they will have a set of assembler arms for building replicas of themselves, an atom or a section at a time.

Speed

How fast these replicators can replicate will depend on their assembly speed and their size. Imagine an advanced assembler that contains a million atoms: it can have as many as ten thousand moving parts, each containing an average of one hundred atoms—enough parts to make up a rather complex machine. In fact, the assembler itself looks like a box supporting a stubby robot arm a hundred atoms long. The box and arm contain devices that move the arm from position to position, and others that change the molecular tools at its tip.

Behind the box sits a device that reads a tape and provides mechanical signals that trigger arm motions and tool changes. In front of the arm sits an unfinished structure. Conveyors bring molecules to the assembler system. Some supply energy to motors that drive the tape reader and arm, and others supply groups of atoms for assembly. Atom by atom (or group by group), the arm moves pieces into place as directed by the tape; chemical reactions bond them to the structure on contact. These assemblers will work fast. A fast enzyme, such as *carbonic anhydrase* or *ketosteroid isomerase*, can process almost a million molecules per second, even without conveyors and power-driven mechanisms to slap a new molecule into place as soon as an old one is released. It might seem too much to expect an assembler to grab a molecule, move it, 'and jam it into place in a mere millionth of a second. But small appendages can move to and fro very swiftly. A human arm can flap up and down several times per second, fingers can tap more rapidly, a fly can wave its wings fast enough to buzz, and a mosquito makes a maddening whine. Insects can wave their wings at about a thousand times the frequency of a human arm be cause an insect's wing is about a thousand times shorter.

An assembler arm will be about fifty million times shorter than a human arm, and so (as it turns out) it will be able to move back and forth about fifty million times more rapidly. For an assembler arm to move a mere million times per second would be like a human arm moving about once per minute: sluggish. So it seems a veiy reason able goal.

The speed of replication will depend also on the total size of the system to be built. Assemblers will not replicate by themselves; they will need materials and energy, and instructions on how to use them. Ordinary chemicals can supply materials and energy, but nanomachinery must be available to process them, Bumpy polymer molecules can code information like a punched paper tape, but a reader must be available to translate the patterns of bumps into pat terns of arm motion. Together, these parts form the essentials of a replicator: the tape supplies instructions for assembling a copy of the assembler, of the reader, of the other nanomachines, and of the tape itself.

A reasonable design for this sort of replicator will likely include several assembler arms and several more arms to hold and move work pieces. Each of these arms will add another million atoms or so. The other parts—tape readers, chemical processors, and so forth—may also be as complicated as assemblers. Finally, a flexible replicator system will probably include a simple computer; following the mechanical approach, this will add roughly 100 million atoms. Altogether, these parts will total less than 150 million atoms. Assume instead a total of one *billion*, to leave a wide margin for error. Ignore the added capability of the additional assembler arms, leaving a still wider margin. Working at one million atoms per second, the system will still copy itself in one thousand seconds, or a bit over fifteen minutes—about the time a bacterium takes to replicate under good conditions.

Imagine such a replicator floating in a bottle of chemicals, making copies of itself. It builds one copy in one thousand seconds, thirty-six in ten hours. In a week, it stacks up enough copies to fill the volume of a human cell. In a century, it stacks up enough to make a respect able speck. If this were all that replicators could do, we could perhaps ignore them in safety. Each copy, though, will build yet more copies. Thus the first replicator assembles a copy in one thousand seconds, the two replicators then build two more in the next thousand seconds, the four build another four, and the eight build another eight. At the end of ten hours, there are not thirty-six new replicators, but over 68 billion. In less than a day, they would weigh a ton; in less than two days, they would outweigh the Earth; in another four hours, they would exceed the mass of the Sun and all the planets combined—if the bottle of chemicals hadn't run dry long before.

Growth

Regular doubling means *exponential growth*. Replicators multiply exponentially unless restrained, as by lack of room or resources. Bacteria do it, and at about the same rate as the replicators just de scribed. People replicate far more slowly, yet given time enough they, too, could overshoot any finite resource supply. Concern about population growth will never lose its importance. Concern about controlling rapid new replicators will soon become important in deed.

Molecules and Skyscrapers

Machines able to grasp and position individual atoms will be able to build almost anything by bonding the right atoms together in the right patterns. To be sure, building large objects one atom at a time will be slow. A fly, after all, contains about a million atoms for every second since the dinosaurs were young. Molecular machines can nonetheless build objects of substantial size—they build whales, after all. To make large objects rapidly, a vast number of assemblers must cooperate, but replicators will produce assemblers by the ton. In deed, with correct design, the difference between an assembler system and a replicator will lie entirely in the assembler's programming. If a replicating assembler can copy itself in one thousand seconds, then it can be programmed to build something else its own size just as fast. Similarly, a ton of replicators can swiftly build a ton of something else—and the product will have all its billions of billions of billions of atoms in the right place, with only a minute fraction misplaced.

To see the abilities and limits of one method for assembling large objects, imagine a flat sheet covered with small assembly arms—perhaps an army of replicator reprogrammed for construction work and arrayed in orderly ranks. Conveyors and communication channels behind them supply reactive molecules, energy, and assembly instructions. If each arm occupies an area 100 atomic diameters wide, then behind each assembler will be room for conveyors and channels totaling about 10,000 atoms in cross-sectional area. This seems room enough. A space ten or twenty atoms wide can hold a conveyor (perhaps based on molecular belts and pulleys). A channel a few atoms wide can hold a molecular rod which, like those in the mechanical computer, will be pushed and pulled to transmit signals. All the arms will work together to build a broad, solid structure layer by layer. Each arm will be responsible for its own area, handling about 10,000 atoms per layer. A sheet of assemblers handling 1,000,000 atoms per second per arm will complete about one hundred atomic layers per second. This may sound fast, but at this rate piling up a paper-sheet thickness will take about an hour, and making a meter-thick slab will take over a year.

Faster arms might raise the assembly speed to over a meter per day, but they would produce more waste heat. If they could build a meter-thick layer in a day, the heat from .one square meter could cook hundreds of steaks simultaneously, and might fry the machinery. At some size and speed, cooling problems will become a limiting factor, but there are other ways of assembling objects faster without overheating the machinery. Imagine trying to build a house by gluing together individual grains of sand. Adding a layer of grains might take grain machines so long that raising the walls would take decades. Now imagine that machines in a factory first glue the grains together to make bricks. The factory can work on many bricks at once. With enough grain machines, bricks would pour Out fast; wall assemblers could then build walls swiftly by stacking the preassembled bricks. Similarly, molecular assemblers will team up with larger assemblers to build big things quickly—machines can be any size from molecular to gigantic. With this approach, most of the assembly heat will be dissipated far from the work site, in making the parts.

Skyscraper construction and the architecture of life suggest a related way to construct large objects. Large plants and animals have vascular systems, intricate channels that carry materials to molecular machinery working throughout their tissues. Similarly, after riggers and riveters finish the frame of a skyscraper, the building's "vascular system"—its elevators and corridors, aided by cranes—carry construction materials to workers throughout the interior. Assembly systems could also employ this strategy, first putting up a scaffold and then working throughout its volume, incorporating materials brought through channels from the outside.

Imagine this approach being used to "grow" a large rocket engine, working inside a vat in an industrial plant. The vat—made of shiny steel, with a glass window for the benefit of visitors—stands taller than a person, since it must hold the completed engine. Pipes and pumps link it to other equipment and to water-cooled heat ex changers. This arrangement lets the operator circulate various fluids through the vat. To begin the process, the operator swings back the top of the vat and lowers into it a base plate on which the engine will be built. The top is then resealed. At the touch of a button, pumps flood the chamber with a thick, milky fluid which submerges the plate and then obscures the window. This fluid flows from another vat in which replicating assemblers have been raised and then reprogrammed by making them copy and spread a new instruction tape (a bit like infecting bacteria with a virus). These new assembler systems, smaller than bacteria, scatter light and make the fluid look milky. Their sheer abundance makes it viscous. At the center of the base plate, deep in the swirling, assembler- laden fluid, sits a "seed." It contains a nanocomputer with stored engine plans, and its surface sports patches to which assemblers stick. When an assembler sticks to it, they plug themselves together and the seed computer transfers instructions to the assembler computer. This new programming tells it where it is in relation to the seed, and directs it to extend its manipulator arms to snag more assemblers. These then plug in and are similarly programmed. Obeying these instructions from the seed (which spread through the expanding net work of communicating assemblers) a sort of assembler-crystal grows from the chaos of the liquid. Since each assembler knows its location in the plan, it snags more assemblers only where more are needed. This forms a pattern less regular and more complex than that of any natural crystal. In the course of a few hours, the assembler scaffolding grows to match the final shape of the planned rocket engine.

Then the vat's pumps return to life, replacing the milky fluid of unattached assemblers with a clear mixture of organic solvents and dissolved substances—including aluminum compounds, oxygen-rich compounds, and compounds to serve as assembler fuel. As the fluid clears, the shape of the rocket engine grows visible through the window, looking like a full-scale model sculpted in translucent white plastic. Next, a message spreading from the seed directs designated assemblers to release their neighbours and fold their arms. They wash out of the structure in sudden streamers of white, leaving a spongy lattice of attached assemblers, now with room enough to work. The engine shape in the vat grows almost transparent, with a hint of iridescence.

Each remaining assembler, though still linked to its neighbours, is now surrounded by tiny fluid-filled channels. Special arms on the assemblers work like flagella, whipping the fluid along to circulate it through the channels. These motions, like all the others performed by the assemblers, are powered by molecular engines fueled by molecules in the fluid. As dissolved sugar powers yeast, so these dissolved chemicals power assemblers. The flowing fluid brings fresh fuel and dissolved raw materials for construction; as it flows out it carries off waste heat. The communications network spreads instructions to each assembler.

The assemblers are now ready to start construction. They are to build a rocket engine, consisting mostly of pipes and pumps. This means building strong, light structures in intricate shapes, some able to stand intense heat, some full of tubes to carry cooling fluid. Where great strength is needed, the assemblers set to work constructing rods of interlocked fibers of carbon, in its diamond form. From these, they build a lattice tailored to stand up to the expected pattern of stress. Where resistance to heat and corrosion is essential (as on many surfaces), they build similar structures of aluminum oxide, in its sapphire form. In places where stress will be low, the assemblers save mass by leaving wider spaces in the lattice. In places where stress will be high, the assemblers reinforce the structure until the remaining passages are barely wide enough for the assemblers to move. Elsewhere the assemblers lay down other materials to make sensors, computers, motors, solenoids, and whatever else is needed.

To finish their jobs, they build walls to divide the remaining channel spaces into almost sealed cells, then withdraw to the last openings and pump out the fluid inside. Sealing the empty cells, they withdraw completely and float away in the circulating fluid. Finally, the vat drains, a spray rinses the engine, the lid lifts, and the finished engine is hoisted out to dry. Its creation has required less than a day and almost no human attention. What is the engine like? Rather than being a massive piece of welded and bolted metal, it is a seamless thing, gemlike. Its empty internal cells, patterned in arrays about a wavelength of light apart, have a side effect: like the pits on a laser disk they diffract light, producing a varied iridescence like that of a fire opal. These empty spaces lighten a structure already made from some of the lightest, strongest materials known. Compared to a modern metal engine, this advanced engine has over 90 percent less mass.

Tap it, and it rings like a bell of surprisingly high pitch for its size. Mounted in a spacecraft of similar construction, it flies from a runway to space and back again with ease. It stands long, hard use because its strong materials have let designers include large safety margins. Be cause assemblers have let designers pattern its structure to yield be fore breaking (blunting cracks and halting their spread), the engine is not only strong but tough.

For all its excellence, this engine is fundamentally quite conventional. It has merely replaced dense metal with carefully tailored structures of light, tightly bonded atoms. The final product contains no nanomachinery.

More advanced designs will exploit nanotechnology more deeply. They could leave a vascular system in place to supply assembler and disassembler systems; these can be programmed to mend worn parts. So long as users supply such an engine with energy and raw materials, it will renew its own structure. More advanced engines can also be literally more flexible. Rocket engines work best if they can take different shapes under different operating conditions, but engineers cannot make bulk metal strong, light, and limber. With nanotechnology, though, a structure stronger than steel and lighter than wood could change shape like muscle (working, like muscle, on the sliding- fiber principle). An engine could then expand, contract, and bend at the base to provide the desired thrust in the desired direction under varying conditions. With properly programmed assemblers and disassemblers, it could even remodel its fundamental structure long after leaving the vat.

In short, replicating assemblers will copy themselves by the ton, then make other products such as computers, rocket engines, chairs, and so forth. They will make disassemblers able to break down rock to supply raw material. They will make solar collectors to supply energy. Though tiny, they will build big. Teams of nanomachines in nature build whales, and seeds replicate machinery and organize atoms into vast structures of cellulose, building redwood trees. There is nothing too startling about growing a rocket engine in a specially prepared vat. Indeed, foresters given suitable assembler "seeds" could grow spaceships from soil, air, and sunlight. Assemblers will be able to make virtually anything from common materials without labour, replacing smoking factories with systems as clean as forests. They will transform technology and the economy at their roots, opening a new world of possibilities. They will indeed be engines of abundance.

8

Industrial Auto-replicator

Imagine a molecule that likes its own shape: finding a copy of itself, it will fit neatly with its twin, forming for a while a complete entity. If the original molecule is presented with the component parts of itself, it will assemble these into additional replicas. The process will continue as long as the supply of components lasts.

Attempts to imitate life are still very young compared to the beginnings of life on the earth perhaps three and a half billion years ago. No one can say for sure what the atmospheric or terrestrial conditions were at the time or which molecule crossed the critical frontier between organic chemistry and biology.

In 1953 Stanley L. Miller, then at the University of Chicago, made one of the first-if not the first-attempts to recreate this transition. He formed amino acids in a mixture of water, methane, ammonia and hydrogen-substances thought to have been present on the primitive earth-by subjecting the chemicals to an electric discharge. James D. Watson and Francis Crick's unraveling of the structure of DNA—also in 1953—further fueled this quest.

For some decades now, the most widely accepted recipe for the origin of life has specified DNA or RNA in lukewarm water, neither acidic nor alkaline, plus only those reagents that were presumably part of the primitive earth's atmosphere. Practitioners of prebiotic chemistry such as Miller and Leslie E. Orgel of the Salk Institute for Biological Studies in San Diego have provided deep insights into molecular replication under these constraints. Indeed, countless metric tons of DNA replicate every day in living creatures in precisely this manner, under the surveillance of a host of enzymes.

But recent findings indicate that the primitive earth was perhaps not quite as hospitable as the "warm pond" paradigm would have [see "In the Beginning...," by John Horgan; Scientific American, February 19911. Such suspicions, coupled with discoveries of organisms living at (literally) bloodcurdling temperatures or near sulfurous volcanic vents at the bottom of the ocean, have led scientists to suggest that life possibly arose by some very different route. Perhaps the only qualities essential to the molecules that become live are detailed in two remarkable books by Richard Dawkins, The Blind Watchmaker and The Selfish Geng. Although written 15 years ago, Dawkins's sketches astonishingly prefigure the results of the past four years of my work on self-replication.

Molecules, natural or synthetic, are able to replicate when their shapes and chemistry have a feature called *complementarity*. By virtue of the way a molecule occupies space and the way its attracting atoms or groups of atoms are distributed along its arms, one molecule may fit snugly into the chemical nooks and crannies of another. The "goodness of fit" between two such complementary molecules thus depends not only on their spatial structure but also on the different kinds of chemical bonds that hold them together in groups. Such groups, or "complexes," form and dissipate rapidly in microseconds or nanoseconds-times that are very short, yet long enough for chemical reactions to take place.

The forces holding complexes together are many times weaker than the covalent bonds binding atoms into molecules. One kind of force-important in complexes is called a *hydrogen bond*. This bond comes about when a hydrogen atom possessing a partial positive charge is attracted to, for instance, an oxygen atom that has a partial negative charge. More general attractions of this class go by the name of polar interactions.

Another kind of force, the van der Waals force, is more subtle: if correctly positioned, electrons of one molecule can jostle away those of another, creating a charge imbalance that results in attraction. Yet a third kind of attraction is "aromatic stacking-an arrangement at flat organic molecules (often having a pleasant odor; hence the name) sometimes assume when they do not like the solvent they find themselves in. By sidling up to one another, flat surface to flat surface, they can squeeze out all the solvent molecules between them and achieve a more stable, stacked configuration.

Once a complex forms, the molecular surfaces that match up with one another are relatively protected. Destructive solvents, dissolved acids, bases or oxidants cannot get to them. Strong covalent bonds then have time to join the complementary parts. Sometimes two of three molecules in a complex link together; the third merely serves to ease the process.

Such a coupling give rise to a rather popular scheme or replication-the one preferred by DNA. A simple depletion of this scheme uses concave and convex shapes. A concave molecular surface lined with appropriately enticing atoms can recognize and surround its convex complement. Further, it can act as a mold for assembling the convex molecule from its component parts. In turn, the convex molecule serves as a template for gathering and fusing the component parts of the concave one. These two replication events-each molecule forming the other-establish what is called a *bicycle*. Our recent experiments indicate that a bicycle can be extremely efficient.

There is an alternative paradigm of replication: two complementary molecules in a complex can join at some site that is not on the recognition surface. They form a single molecule, one end of which is complementary to the other and the whole is complementary to itself. The recognition surfaces at the ends of this new, self-complementary molecule are still accessible to other molecules. The ends can each gather a fragment identical to that at the other end.

Once gathered, the two new components cannot move freely and travel through space in tandem; the chances of their becoming linked to each other are greatly enhanced. Thus, the self complementary entity makes a copy and in similar manner, many copies of itself. No enzymes are needed: the molecule catalyzes its own formation.

This is the method we have used in the laboratory to make molecules capable of reacting with one another in ways reminiscent of life. Among them are molecules that bear a passing resemblance to genetic materials-specifically, to nucleic acid components known as adenines. Adenines are flat; besides, they have hydrogen and nitrogen atoms that can form hydrogen bonds with the oxygen and hydrogen atoms of their complementary molecules, called imides. Our imides are constructed from a humpbacked molecule, Kemp's triacid, the skeleton of which folds over in such a way that large, concave structures can easily be fashioned from it. So the imide features a hydrogen bonding site crookedly attached to an aromatic stacking surface; these fit perfectly with the hydrogen bonding site and the flat stacking surface of adenine. When associated together in a complex, the adenine and the imide become covalently attached, forming a self-complementary molecule. Our early attempts to get this molecule to self-replicate were thwarted by its unforeseen floppiness. Although some flexibility is helpful for molecular recognition—a leather boot is easier to slip on than a wooden one—a lot of flexibility can make fitting very difficult—try slipping on a sock without using your hands. Molecules become floppy if they have single bonds, involving only two electrons each. Such a bond allows the parts it joins to rotate with respect to each other, giving rise to many different shapes.

This new J-shaped molecule, adenine ribose naphthalene imide (ARNI for, short), provided us with our first instance of replication. Using high-performance liquid chromatography to detect minute changes in chemical concentrations, Tjivikua and Pablo Ballester, a postdoctoral visitor from Majorca, achieved the result. They compared the rate of formation of ARNI in a solution that contained only its components with the rate of formation when some ARNI was added. The presence of ARNI increased the rate of formation, clear evidence of the presence of a self-replicating system.

If one plots the progress of a reaction through time, one generally derives a curve that assumes the shape of a reclining parabola. The product forms fastest at the beginning, when the reactants are at their highest concentrations; the rate of formation slows down as the reactants are consumed. For an autocatalytic reaction one in % rich the product, like our ARNI, catalyzes its own formation-the growth curve should be S-shaped, or "sigmoidal". The reaction begins slowly. As the product appears and begins to act as a catalyst, the reaction accelerates. An upward curve results. Finally, as the materials are consumed, the reaction grinds to a halt.

The degree of sigmoidal curvature depends on several factors, the most important of which is the efficiency of the autocatalytic step. If the background reaction-in which the components combine by themselves, without getting help from the self-replicating molecule-is too strong, it can swamp the signal from the self-replicating process. In 1990 Gilnther von Kiedrowski and his co-workers in Gottingen showed that a self-replicating nucleic acid could exhibit such sigmoidal growth-proving that the autocatalyzed synthesis is in this case more efficient than the random one.

Although ARNI did not show sigmoidal growth, our next attempt, ARBI, did. We slowed the background reaction rates by giving ARBI a slightly longer stacking element, a biphenyl instead of naphthalene. We now had proof of a bonafide synthetic self-replicating molecule. Is it alive? Not by most current definitions. Our (or its) critics were quick to point out that as a life-form, ARBI had severe limitations: the molecule would make copies only of itself. To allow evolution, a self-replicating molecule has to be capable of "making mistakes": occasionally synthesizing other molecules that can perhaps be better replicators. Unlike art and music critics, those in science at least indicate in which direction improvements may he. We responded by devising molecules that were capable of making -indeed, that were incapable of not making mistakes.

A molecule can readily be manipulated into making replicas of a competitor. Instilling the capacity to "evolve" takes more planning. In the course of our search for a solution, we tapped into some earlier findings on the hydrogen bonding sites of adenine. There are two ways in which an imide can attach to an adenine. It can find a site along the "Watson-Crick edge, which is involved in the replication of DNA, or it can dock along the Hoogsteen edge, a region normally exposed in DNA (though sometimes joined in such exotic forms as triple helices).

We had already shown that simple adenines can attach to our imides along either edge. For example, roughly equal amounts of Watson-Crick and Hoogsteen complexes are formed with an imide attached to a naphthalene surface. But if one of the hydrogens of the amino (NHZ) group of adenine-involved in hydrogen bonding-is replaced by a larger group of molecules, the situation changes. This new group positions itself in such a way as to block access to the Watson-Crick edge while leaving the Hoogsteen edge largely open. For example, when a small methyl group joins the adenine, more than 85 percent of our synthetic imide receptors are found to bind along the Hoogsteen edge.

Now came the challenge of demonstrating a chemical version of mutation: a permanent, heritable structural change affecting the survival capacity of an organism or in our case its analogue, a self-replicating

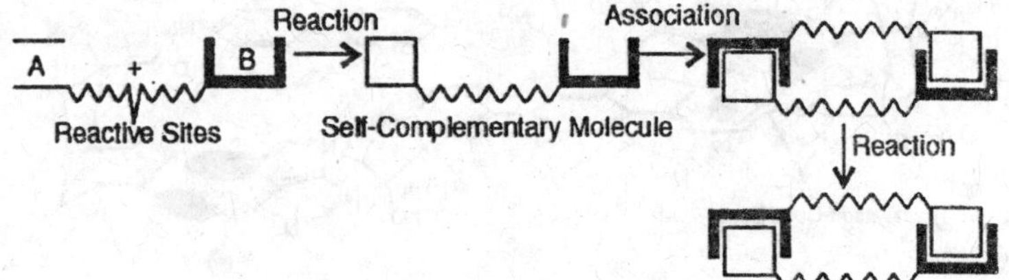

Fig. 8.1. Self-replication.

molecule. Changes in a molecular replicator's structure might be caused by shifts in temperature, acidity, salinity or many other factors. We chose irradiation with light.

We first placed ZARBI and ZNARBI in direct competition for reproductive resources. Qing Feng, then my graduate student, and Jong-in Hong, a postdoctoral fellow, allowed the adenine derivatives bearing the Z and Z-NO2 groups to compete for a limited amount of the complementary biphenyl receptor. ZNARBI proved to be a slightly better replicator. When all the receptor was used up, we irradiated the reaction vessel with ultraviolet light having a wavelength of 350 nanometers. After a few hours of irradiation, the Z-NO2 blocking groups had all been removed from both the ZNARBI replicators as well as from the adenine progenitors. That is, the ZNARBI molecules had all been converted to ARBIs and the Z-NO2 bearing adenines to

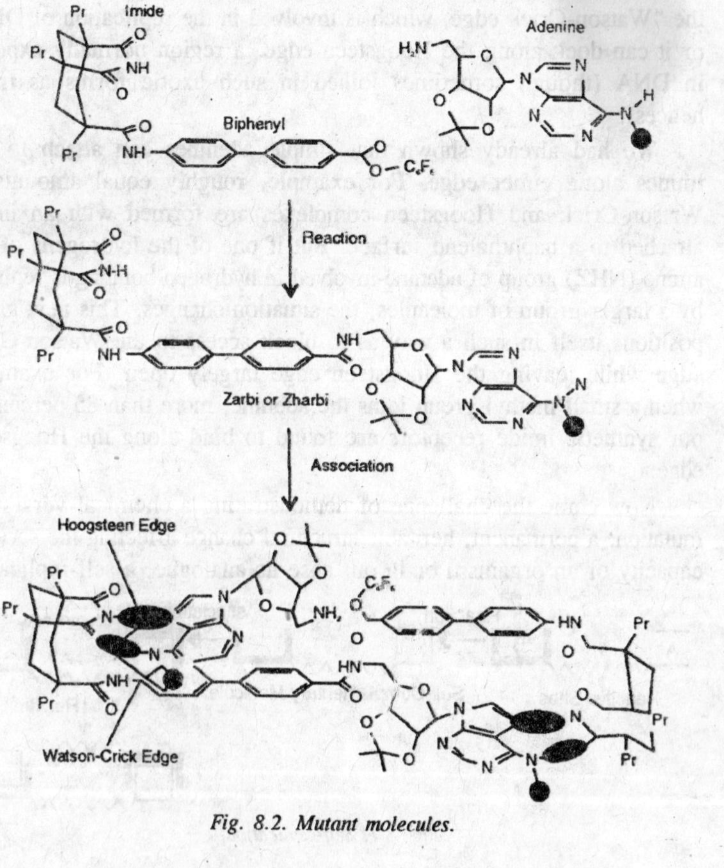

Imide
Adenine
Biphenyl
Reaction
Zarbi or Zharbi
Association
Hoogsteen Edge
Watson-Crick Edge

Fig. 8.2. Mutant molecules.

plain adenines. A "mutation" had occurred, prompted by a change in the environment. ZARBI and the Z-adenine remained unaltered.

Next we added more biphenyl receptor. ZARBI found the radiation product, ARBI, to be its competitor. The sleeker ARBI, having the advantage of replicating in either the Hoogsteen or the Watson-Crick mode, rapidly took over the resources of the system.

A simple evolutionary interpretation can be sketched for this experiment. Think of ZARBI as the original in this sequence; its replication requires the presence of Z-adenine and the biphenyl receptor. If nitric acid is added, then some of the Z-adenines are used up to form Z-NO2 adenines; the latter give rise to ZNARBIs. ZNARBI is a better replicator compared with its ancestor ZARBI. When the system is irradiated, a second change takes place. ZNARBI converts to the simpler and more efficient ARBI. This last molecule proves to be the best replicator of the three.

Although mutation is considered, to be the driving force for most evolutionary alterations, another significant paradigm for change is recombination. Two chromosomes can split, exchange strings of DNA and rejoin, thus combining their characteristics. Also, certain computer programs attempt to "teach" strings of information to solve a problem. If the strings are allowed to split and recombine at random, they soon give rise to much better problem solvers. Mutation allows for single, small changes; recombination, on the other hand, allows the creation of hybrids that are very different from the progenitors.

How did this difference in the ability to replicate come about? The efficiency of the ART replicator is easily rationalized. ART looks a good deal like a piece of DNA, possibly the best replicator in existence. Furthermore, its ribose piece makes the recognition surfaces parallel to one another, a very helpful configuration. This and the high affinity of adenine for its complement thymine make for an easily assembled complex-the intermediate stage in replication.

The inefficiency of DIXBI can also be traced to its overall molecular shape. DIXBI is composed of two U-shaped molecules connected by a rigid biphenyl spacer; its overall structure can adopt a C or an S shape. The recognition surfaces are exposed inside the C shape, where there is not enough room for a replicating complex to form. In the S shape the recognition surfaces are far apart, so that when a complex forms, the reactive pieces are too far away from one another to bond covalently. Thus, even though DIXBI is self-complementary, it cannot achieve replication.

There are good indications that a smaller molecule, such as a solvent chloroform molecule, can fit within our molecular tennis ball. But it is too small to accommodate even our most minimal replicators. We are now working with Javier de Mendoza of the Free University of Madrid to develop a larger molecule—a softball—that may have an interior roomy enough to hold some of our replicators.

Once we have made it past the problem of containment, the biggest obstacle to the molecular life agenda will be these questions: How can our fledgling organism harness energy? From sunlight or from other molecules? How can the component pieces of the replicators and their containers be replenished? These are the challenges of the next decade. Whether they are met or not, the efforts of chemists to answer them will surely provide insight into the organic chemistry of life-how it came about and how it continues to flourish.

BACKGROUND

Long a fascination of science fiction writers and space exploration visionaries, nanomachines that can replicate themselves are already here: we have an existence proof on Earth in complex living systems, such as one-celled (and even more complex many-celled) organisms. Other "bionanomachines" viruses are able to reproduce themselves through the use of cell machinery external to themselves. Creating synthetic self-replicators would greatly scale up production of nanomachines from the atomic and molecular scale to the macroworld, as the process of self-replication allows for exponential growth. Your task is to propose a scientific plan for the design and creation of a simplified synthetic self-replicating nanomachine, using a replication method either completely self-contained, as in a cell, or requiring the use of external machinery, such as by a virus.

Problem

1. All cells on earth appear to be built according to the same molecular plan, using evolved molecular self-replication: Ribosomes are molecular assemblers working from stored information.
2. DNA is the information storage medium, directing the assembly of other parts.
3. DNA polymerase duplicates the information storage medium.
4. Thousands of enzymes convert available raw materials to building blocks required for the assembler and duplicator.
5. The cell provides the following required infrastructure: (i) The lipid cell membrane serves to define the body of the cell. (ii) The

membrane signaling and transport proteins serve to allow for communication, energy, and raw material transport to and from the external environment. (iii) Complex machinery exists to allow for reproduction by binary fission. (iv) Various enzymes exist for regulation and error correction of cell processes.

6. Current estimates of the minimal number of DNA genes needed to create a living organism modeled after the modern cell machinery above are in the range of 250-350; however, this design is limited by the process of evolution. Can we design a more efficient and simpler self-replicator? For example, it is believed that primordial life was based on RNA, and there are attempts to create RNA ribozymes in the lab as well as a major advance in understanding of ancient RNA processes that still exist in modern organisms. Another possibility would be to create stable alternatives to DNA and RNA, such as synthetic short peptide chains that can be more robust for information storage and control. Yet another possibility is to create self-replicating DNA objects using synthetic DNA structures as engineering materials akin to viruses, requiring access to external machinery for replication.
7. Current cell machinery is limited to water environments and thus a limited temperature range, in which thermal statistical motion and a diffusion-to-capture paradigm occurs for most functional tasks. Larger and more specialized tasks are carried out by machine-phase assemblies. Could we design self-replicators that evolve and grow in environments without water?
8. DNA, RNA, and most proteins have limited lifetimes in cells due to degradation by nucleases and denaturization. Is it possible to create more robust and longer lived replicators? What are the trade-offs?
9. The measured mutation rate in bacterial cells is 1 nucleotide in 10^9 nucleotide polymerization events. What level and kinds of transcription and replication error rate and error correction processes are needed to sustain self-replicating nanomachines? Although transcription errors can be fatal, some types of transcription errors, along with gene duplication and complex gene networks, can help an organism evolve in a changing environment.
10. Self-replicating nanomachines would have many positive uses for society, but their possible existence in the near future also raises many concerns of "gray goo" either inadvertently or purposefully being unleashed on the environment with unforeseen possible grave

consequences. What kind of ethical controls should be put in place over their creation and use?

Solution

Focus Group 2 met to discuss how scientists might develop synthetic self-replicators, devices that can make copies of themselves. These devices could have many valuable applications. Substances made of microscopic self-replicators could heal themselves by producing replacements for damaged parts. For example, molecular scale self-replicators could combine to form self-maintaining paint or spacecraft skins that can repair damage caused by space debris. In addition to replacing damaged parts, self-replicators can scale up production exponentially, as each new product is at the same time a new factory for more products. This could be a solution for accurately and inexpensively producing useful quantities of novel nanoscale materials.

While technological applications have caught the attention of many, including science fiction writers, researchers are also excited about potential non-technological payoffs for research into self-replicators. Building our own self-replicators could give us insight into the origins and mechanisms of existing self-replicators, ranging from bacteria to balboa trees, all of life, in fact, including ourselves.

Life serves as proof that self-replication is in fact possible. Other, non-living self-replicators also exist. Because a range of self-replicators exist in the world, the members of the focus group had to first define the parameters for designing a model self-replicating system.

The group discussed several types of existing self-replicators. First, several examples of simple replicators were named. Fire, in the right environment, produces more fire. Crystal seeding leads to more crystal. Autocatalytic reactions produce chemicals that in turn increase the reaction. For example, if hit with a source of energy, like gamma rays, formaldehyde makes glycoaldehyde. Once glycoaldehyde is present, it can couple with formaldehyde and break it apart, making two glycoaldehyde molecules where there had been one. These in turn can convert more formaldehyde to glycoaldehyde. As long as formaldehyde is available, this reaction causes more of itself to occur.

Viruses fall into another category of self-replicators. They are more complex than fire, but to make copies of themselves they have to depend upon the machinery inside biological cells. One of the things that make viruses interesting is that, like life, they carry instructions for copying themselves. They inject either DNA or RNA into a cell, where cellular machinery follows the directions and produces more viruses.

The last category of self-replicators the group considered was biological cells. In part because many in the group hoped to use the pursuit of a synthetic self-replicator to throw light on the origins of life, the group decided to specify a self-replicator much like a cell. Like fire and crystals, its self-replicator would make copies of itself. Like viruses, it would contain instructions for self-replication. It would be like a cell in many ways. First, unlike viruses, the replicator would include the machinery for carrying out the instructions. Also, it would take simple environmental materials, as cells use amino acids, and create something more complex, such as a cell's proteins. The group wanted to make clear it was not looking for a self-replicator that made copies of itself by, for example, breaking off parts of a more complex material in the environment.

In addition to these basic requirements, the group hoped its self-replicator would have other things in common with a cell. The instructions in a cell can be changed, and as a result the cell can produce different kinds of products and perform various functions. Muscle cells can contract. Nerve cells can process and send signals. Likewise, an ideal synthetic self-replicator would be programmable so that it could serve multiple functions.

The group decided its self-replicator could be different than a cell in one important way: it would not necessarily have to have a physical barrier like the cell's membrane. The group's self-replicator still would need to be distinct from its environment, if only to confirm that it is indeed making a copy of itself. Rather than using a physical barrier, however, this distinction could be made by defining the parts or functions of the self-replicator.

By agreeing not to include a requirement for a cell membrane-like physical boundary, the group significantly reduced the complexity of the design task. At the same time, the group increased the requirement for researchers to control the environment for the self-replicator. In a cell, the membrane, including its embedded proteins, control what comes into the cell. By doing this it creates a special environment within the cell that allows the reactions necessary for the cell to function and eventually copy itself. For the group's self-replicator, the researchers in effect take the place of the membrane, carefully preparing and maintaining the environment. They would keep out things that might damage the machine, and they would include an energy source and all the required raw materials. The need for this specified environment makes it much less likely that this self-replicator could survive and reproduce outside of the lab.

In summary, the group defined as its goal a self-replicator that:

- produces a copy of itself
- carries information for replication
- is distinct from its environment
- uses raw materials that are simpler than the final product
- ideally would be programmable and multifunctional

The group's defined goal will not be easy to accomplish. As a first step, however, the group outlined a research direction building on current work with RNA. David Bartel of the Massachusetts Institute of Technology has developed an RNA-based RNA polymerase, that is, a form of RNA that can copy RNA. If this polymerase could make a copy of its own RNA sequence, it would be a self-replicator.

For this to happen, key obstacles need to be overcome. For one thing, so far the polymerase is slow and as a result cannot copy long strands of RNA such as itself. Another main problem is the fact that once the RNA is copied and folds into a non-linear structure, like a helix, its parts are no longer available to be copied again. What is needed is another enzyme, a helicase, that will unfold the structure so it can be copied.

In spite of these obstacles, working with RNA seems promising because, in addition to possibly fulfilling the group's basic requirements, it might even lead to a device that can be programmed to perform a variety of functions. Nucleic acids have been used for a variety of surprising things. Researchers have made DNA that folds into an octahedron, opens and closes like a pair of tweezers, or walks on a substrate much as the protein molecular motor kinesin walks along microtubules. They have also used RNA for a variety of catalytic roles. Even more functions may be found if the so-called RNA world hypothesis is correct. According to Nobel Prize for Chemistry winner Sidney Altman, in the primitive earth RNA both stored genetic information and performed, "the full range of catalytic roles necessary in a very primitive self-replicating system." If scientists are able to synthesize an RNA-based self-replicator, it may confirm this hypothesis and give us a better understanding of how life could have begun and evolved.

The proposed self-replicator might work something like this: RNA polymerase would be added to a solution containing all the raw materials it needs, including fuel in the form of rNTP. The helicase would unfold some of them, making them available for copying by other, still folded, molecules of RNA polymerase. These copies would fold into new RNA polymerase molecules. These could be fed other strands of RNA that code for RNA-based structures like tweezers and catalysts, or more polymerase.

After offering the RNA example, the group went on to suggest that non-biological heteropolymers might be used to make self-replicating machines that could survive within extreme environments like space, where the vacuum, cold, and radiation would keep biological self-replicators from functioning or even maintaining integrity. Such non-biological self replicators would depend upon a supply of raw materials that do not occur naturally, suggesting that they would not be able to replicate outside of a carefully prepared environment. While the theoretical advantages of non-biological heteropolymers make them desirable, the group noted that building them would present an array of new obstacles.

Public concerns about self-replicators have been heightened by books like Michael Crichton's *Prey*. Although the replicators proposed by the group would likely have trouble surviving outside of narrow environments, the group proposed that attempts to make self-replicators should be accompanied by critical assessments of safety issues, including consideration of ways to recognize and respond to unforeseen problems. These assessments from the beginning should include discussions between scientists and nonscientists with the goal of self-regulation.

9

NANOELECTRONICS

Instrumentation and measurement technologies are some of the most promising areas for near-term advancements and enabling effects. As optical, fluidic, chemical, and biological components can be integrated with electronic logic and memory components on the same chip at marginal... to perform basic diagnostic functions in a minimally invasive way, providing new abilities to remedy health problems.

Other possible applications include: pervasive, self-moving sensor systems; nanoscrubbers and nanocatalysts; even inexpensive, networked "nanosatellites." For example, so-called "nanosatellites" are targeting order-of-magnitude reductions in... proach by Prentiss uses lasers to place molecules in a desired location. Chemical assembly techniques are being addressed by a number of groups, including Whiteside's approach to building structures one molecular layer at a time.

A third major area for development within molecular manufacturing is systems design and engineering....understanding of fundamental properties of structures at the nanoscale has been greatly enhanced by the ability to fabricate very small test objects, analyze them experimentally using new capabilities.

MICRO/NANOSCALE INSTRUMENTATION

Coal and diamonds, sand and computer chips, cancer and healthy tissue: throughout history, variations in the arrangement of atoms have distinguished the cheap from the cherished, the diseased from the healthy. Arranged one way, atoms make up soil, air, and water; arranged another, they make up ripe strawberries. Arranged one way, they make up homes and fresh air; arranged another, they make up ash and smoke. Our ability to arrange atoms lies at the foundation of technology. We have come far in our atom arranging, from chipping flint for arrowheads to machining aluminum for spaceships. We take pride in our technology, with our lifesaving drugs and desktop computers. Yet our spacecraft are still crude, our computers are still stupid, and the molecules in our tissues still slide into disorder, first destroying health, then life itself. For all our advances in arranging atoms, we still use primitive methods. With our present technology, we are still forced to handle atoms in unruly herds.

But the laws of nature leave plenty of room for progress, and the pressures of world competition are even now pushing us forward. For better or for worse, the greatest technological breakthrough in history is still to come.

Styles of Technology

Our modern technology builds on an ancient tradition. Thirty thousand years ago, chipping flint was the high technology of the day. Our ancestors grasped stones containing trillions of trillions of atoms and removed chips containing billions of trillions of atoms to make their axheads; they made fine work with skills difficult to imitate today. They also made patterns on cave walls in France with sprayed paint, using their hands as stencils. Later they made pots by baking clay, then bronze by cooking rocks. They shaped bronze by pounding it. They made iron, then steel, and shaped it by heating, pounding, and removing chips.

We now cook up pure ceramics and stronger steels, but we still shape them by pounding, chipping, and so forth. We cook up pure silicon, saw it into slices, and make patterns on its surface using tiny stencils and sprays of light. We call the products "chips" and we consider them exquisitely small, at least in comparison to axheads.

Our microelectronic technology has managed to stuff machines as powerful as the room-sized computers of the early 1950s onto a few silicon chips in a pocket-sized computer. Engineers are now making ever smaller devices, slinging herds of atoms at a crystal surface to build up wires and components one tenth the width of a fine hair.

These microcircuits may be small by the standards of flint chippers, but each transistor still holds trillions of atoms, and so-called "microcomputers" are still visible to the naked eye. By the standards of a newer, more powerful technology they will seem gargantuan. The ancient style of technology that led from flint chips to silicon chips handles atoms and molecules in bulk; call it bulk technology. The new technology will handle individual atoms and molecules with control and precision; call it *molecular technology*. It will change our world in more ways than we can imagine. *Microcircuits* have parts measured in *micrometers*—that is, in millionths of a meter—but molecules are measured in *nanometers* (a thousand times smaller). We can use the terms "nanotechnology" and molecular technology" interchangeably to describe the new style of technology. The engineers of the new technology will build both nanocircuits and nanomachines.

Molecular Technology Today

One dictionary definition of a machine is "any system, usually of rigid bodies, formed and connected to alter, transmit, and direct applied forces in a predetermined manner to accomplish a specific objective, such as the performance of useful work." Molecular machines fit this definition quite well. To imagine these machines, one must first picture molecules. We can picture atoms as beads and molecules as clumps of beads, like a child's beads linked by snaps. In fact, chemists do sometimes visualize molecules by building models from plastic beads (some of which link in several directions, like the hubs in a Tinkertoy set). Atoms are rounded like beads, and although molecular bonds are not snaps, our picture at least captures the essential notion that bonds can be broken and reformed. If an atom were the size of a small marble, a fairly complex molecule would be the size of your fist. This makes a useful mental image, but atoms are really about 1/10,000 the size of bacteria, and bacteria are about 1/10,000 the size of mosquitoes. (An atomic nucleus, however, is about 1/100,000 the size of the atom itself; the difference between an atom and its nucleus is the difference between a fire and a nuclear reaction.)

The things around us act as they do because of the way their molecules behave. Air holds neither its shape nor its volume because its molecules move freely, bumping and ricocheting through open space. Water molecules stick together as they move about, so water holds a constant volume as it changes shape. Copper holds its shape because its atoms stick together in regular patterns; we can bend it and hammer it because its atoms can slip over one another while remaining bound together. Glass shatters when we hammer it because its atoms separate before they slip. Rubber consists of networks of kinked molecules, like a tangle of springs. When stretched and released, its molecules straighten and then coil again. These simple molecular patterns make up passive substances. More complex patterns make up the active nanomachines of living cells. Biochemists already work with these machines, which are chiefly made of protein, the main engineering material of living cells. These molecular machines have relatively few atoms, and so they have lumpy surfaces, like objects made by gluing together a handful of small marbles. Also, many pairs of atoms are linked by bonds that can bend or rotate, and so protein machines are unusually flexible. But like all machines, they have parts of different shapes and sizes that do useful work. All machines use clumps of atoms as parts. Protein machines simply use very small clumps. Biochemists dream of designing and building such devices, but there are difficulties to be overcome. Engineers use beams of light to project patterns onto silicon chips, but chemists must build much more indirectly than that. When they combine molecules in various sequences, they have only limited control over how the molecules join. When biochemists need complex molecular machines, they still have to borrow them from cells. Nevertheless, advanced molecular machines will eventually let them build nanocircuits and nanomachines as easily and directly as engineers now build microcircuits or washing machines. Then progress will become swift and dramatic. Genetic engineers are already showing the way.

Ordinarily, when chemists make molecular chains—called "*polymers*" —they dump molecules into a vessel where they bump and snap together haphazardly in a liquid. The resulting chains have varying lengths, and the molecules are strung together in no particular order. But in modern gene synthesis machines, genetic engineers build more orderly polymers—specific DNA molecules—by combining molecules in a particular order. These molecules are the nucleotides of DNA (the letters of the genetic alphabet) and genetic engineers don't dump them all in together. Instead, they direct the machine to add different nucleotides in a particular sequence to spell out a particular message. They first bond one kind of nucleotide to the chain ends, then wash away the leftover material and add chemicals to prepare the chain ends to bond the next nucleotide. They grow chains as they bond on nucleotides, one at a time, in a programmed sequence. They anchor the very first nucleotide in each chain to a solid surface to keep the chain from washing away with its chemical bathwater. In this way, they have a big clumsy machine in a cabinet assemble specific molecular structures from parts a hundred million times smaller than itself.

But this blind assembly process accidentally omits nucleotides from some chains. The likelihood of mistakes grows as chains grow longer.

Like workers discarding bad parts before assembling a car, genetic engineers reduce errors by discarding bad chains. Then, to join these short chains into working genes (typically thousands of nucleotides long), they turn to molecular machines found in bacteria. These protein machines, called *restriction enzymes*, "read" certain DNA sequences as "cut here." They read these genetic patterns by touch, by sticking to them, and they cut the chain by rearranging a few atoms. Other enzymes splice pieces together, reading matching parts as "glue here"—likewise "reading" chains by selective stickiness and splicing chains by rearranging a few atoms.

By using gene machines to write, and restriction enzymes to cut and paste, genetic engineers can write and edit whatever DNA messages they choose. But by itself, DNA is a fairly worthless molecule. It is neither strong like Kevlar, nor colourful like a dye, nor active like an enzyme, yet it has something that industry is prepared to spend millions of dollars to use: the ability to direct molecular machines called ribosomes. In cells, molecular machines first transcribe DNA, copying its information to make RNA "tapes." Then, much as old numerically controlled machines shape metal based on instructions stored on tape, ribosomes build proteins based on instructions stored on RNA strands. And proteins are useful. Proteins, like DNA, resemble strings of lumpy beads. But unlike DNA, protein molecules fold up to form small objects able to do things. Some are enzymes, machines that build up and tear down molecules (and copy DNA, transcribe it, and build other proteins in the cycle of life). Other proteins are hormones, binding to yet other proteins to signal cells to change their behaviour.

Genetic engineers can produce these objects cheaply by directing the cheap and efficient molecular machinery inside living organisms to do the work. Whereas engineers running a chemical plant must work with vats of reacting chemicals (which often misarrange atoms and make noxious by-products), engineers working with bacteria can make them absorb chemicals, carefully rearrange the atoms, and store a product or release it into the fluid around them. Genetic engineers have now programmed bacteria to make proteins ranging from human growth hormone to rennin, an enzyme used in making cheese. The pharmaceutical company Eli Lilly (Indianapolis) is now marketing Humulin, human insulin molecules made by bacteria.

Existing Protein Machines

These protein hormones and enzymes selectively stick to other molecules. An enzyme changes its target's structure, then moves on; a hormone affects its target's behaviour only so long as both remain stuck together. Enzymes and hormones can be described in mechanical terms, but their behaviour is more often described in chemical terms. But other proteins serve basic mechanical functions. Some push and pull, some act as cords or struts, and parts of some molecules make excellent bearings. The machinery of muscle, for instance, has gangs of proteins that reach, grab a "rope" (also made of protein), pull it, then reach out again for a fresh grip; whenever you move, you use these machines. Amoebas and human cells move and change shape by using fibers and rods that act as molecular muscles and bones.

A reversible, variable-speed motor drives bacteria through water by turning a corkscrew-shaped propeller. If a hobbyist could build tiny cars around such motors, several billions of billions would fit in a pocket, and 150-lane freeways could be built through your finest capillaries. Simple molecular devices combine to form systems resembling industrial machines. In the 1950s engineers developed machine tools that cut metal under the control of a punched paper tape. A century and a half earlier Joseph-Marie Jacquard had built a loom that wove complex patterns under the control of a chain of punched cards. Yet over three billion years before Jacquard, cells had developed the machinery of the ribosome. Ribosomes are proof that nanomachines built of protein and RNA can be programmed to build complex molecules.

Then consider viruses. One kind, the T4 phage, acts like a spring-loaded syringe and looks like something out of an industrial parts catalog. It can stick to a bacterium, punch a hole, and inject viral DNA (yes, even bacteria suffer infections). Like a conqueror seizing factories to build more tanks, this DNA then directs the cell's machines to build more viral DNA and syringes. Like all organisms, these viruses exist because they are fairly stable and are good at getting copies of themselves made. Whether in cells or not, nanomachines obey the universal laws of nature. Ordinary chemical bonds hold their atoms together, and ordinary chemical reactions (guided by other nanomachines) assemble them. Protein molecules can even join to form machines without special help, driven only by thermal agitation and chemical forces. By mixing viral proteins (and the DNA they serve) in a test tube, molecular biologists have assembled working T4 viruses. This ability is surprising: imagine putting automotive parts in a large box, shaking it, and finding an assembled car when you look inside! Yet the T4 virus is but one of many self-assembling structures.

Molecular biologists have taken the machinery of the ribosome apart into over fifty separate protein and RNA molecules, and then combined them in test tubes to form working ribosomes again.

To see how this happens, imagine different T4 protein chains floating around in water. Each kind folds up to form a lump with distinctive bumps and hollows, covered by distinctive patterns of oiliness, wetness, and electric charge. Picture them wandering and tumbling, jostled by the thermal vibrations of the surrounding water molecules. From time to time two bounce together, then bounce apart. Sometimes, though, two bounce together and fit, bumps in hollows, with sticky patches matching; they then pull together and stick. In this way protein adds to protein to make sections of the virus, and sections assemble to form the whole. Protein engineers will not need nanoarms and nanohands to assemble complex nanomachines. Still, tiny manipulators will be useful and they will be built. Just as today's engineers build machinery as complex as player pianos and robot arms from ordinary motors, bearings, and moving parts, so tomorrow's biochemists will be able to use protein molecules as motors, bearings, and moving parts to build robot arms which will themselves be able to handle individual molecules.

Designing With Protein

How far off is such an ability? Steps have been taken, but much work remains to be done. Biochemists have already mapped the structures of many proteins. With gene machines to help write DNA tapes, they can direct cells to build any protein they can design. But they still don't know how to design chains that will fold up to make proteins of the right shape and function. The forces that fold proteins are weak, and the number of plausible ways a protein might fold is astronomical, so designing a large protein from scratch isn't easy. The forces that stick proteins together to form complex machines are the same ones that fold the protein chains in the first place. The differing shapes and kinds of stickiness of amino acids—the lumpy molecular "beads" forming protein chains—make each protein chain fold up in a specific way to form an object of a particular shape. Biochemists have learned rules that suggest how an amino acid chain might fold, but the rules aren't very firm. Trying to predict how a chain will fold is like trying to work a jigsaw puzzle, but a puzzle with no pattern printed on its pieces to show when the fit is correct, and with pieces that seem to fit together about as well (or as badly) in many different ways, all but one of them wrong. False starts could consume many lifetimes, and a correct answer might not even be recognized. Biochemists using the best computer programs now available still cannot predict how a long, natural protein chain will actually fold, and some of them have despaired of designing protein molecules soon.

Yet most biochemists work as scientists, not as engineers. They work at predicting how natural proteins will fold, not at designing proteins that will fold predictably. These tasks may sound similar, but they differ greatly: the first is a scientific challenge, the second is an engineering challenge. Why should natural proteins fold in a way that scientists will find easy to predict? All that nature requires is that they in fact fold correctly, not that they fold in a way obvious to people. Proteins could be designed from the start with the goal of making their folding more predictable. Carl Pabo, writing in the journal Nature, has suggested a design strategy based on this insight, and some biochemical engineers have designed and built short chains of a few dozen pieces that fold and nestle onto the surfaces of other molecules as planned. They have designed from scratch a protein with properties like those of melittin, a toxin in bee venom. They have modified existing enzymes, changing their behaviours in predictable ways. Our understanding of proteins is growing daily. In 1959, according to biologist Garrett Hardin, some geneticists called genetic engineering impossible; today, it is an industry.

Biochemistry and computer-aided design are now exploding fields, and as Frederick Blattner wrote in the journal Science, "computer chess programs have already reached the level below the grand master. Perhaps the solution to the protein-folding problem is nearer than we think." William Rastetter of Genentech, writing in *Applied Biochemistry and Biotechnology* asks, "How far off is de novo enzyme design and synthesis? Ten, fifteen years?" He answers, "Perhaps not that long." Forrest Carter of the U.S. Naval Research Laboratory, Ari Aviram and Philip Seiden of IBM, Kevin Ulmer of Genex Corporation, and other researchers in university and industrial laboratories around the globe have already begun theoretical work and experiments aimed at developing molecular switches, memory devices, and other structures that could be incorporated into a protein based computer. The U.S. Naval Research Laboratory has held two international workshops on molecular electronic devices, and a meeting sponsored by the U.S. National Science Foundation has recommended support for basic research aimed at developing molecular computers. Japan has reportedly begun a multimillion-dollar program aimed at developing self-assembling molecular motors and computers, and VLSI Research Inc., of San Jose, reports that "It looks like the race to biochips [another term for molecular electronic systems] has already started.

NEC, Hitachi, Toshiba, Matsushita, Fujitsu, Sanyo-Denki and Sharp have commenced full-scale research efforts on bio-chips for bio-computers."

Biochemists have other reasons to want to learn the art of protein design. New enzymes promise to perform dirty, expensive chemical processes more cheaply and cleanly, and novel proteins will offer a whole new spectrum of tools to biotechnologists.

Second Generation Nanotechnology

Despite its versatility, protein has shortcomings as an engineering material. Protein machines quit when dried, freeze when chilled, and cook when heated. We do not build machines of flesh, hair, and gelatin; over the centuries, we have learned to use our hands of flesh and bone to build machines of wood, ceramic, steel, and plastic. We will do likewise in the future. We will use protein machines to build nanomachines of tougher stuff than protein. As nanotechnology moves beyond reliance on proteins, it will grow more ordinary from an engineer's point of view. Molecules will be assembled like the components of an erector set, and well-bonded parts will stay put. Just as ordinary tools can build ordinary machines from parts, so molecular tools will bond molecules together to make tiny gears, motors, levers, and casings, and assemble them to make complex machines.

Parts containing only a few atoms will be lumpy, but engineers can work with lumpy parts if they have smooth bearings to support them. Conveniently enough, some bonds between atoms make fine bearings; a part can be mounted by means of a single chemical bond that will let it turn freely and smoothly. Since a bearing can be made using only two atoms (and since moving parts need have only a few atoms), nanomachines can indeed have mechanical components of molecular size. How will these better machines be built? Over the years, engineers have used technology to improve technology. They have used metal tools to shape metal into better tools, and computers to design and program better computers. They will likewise use protein nanomachines to build better nanomachines. Enzymes show the way: they assemble large molecules by "grabbing" small molecules from the water around them, then holding them together so that a bond forms. Enzymes assemble DNA, RNA, proteins, fats, hormones, and chlorophyll in this way—indeed, virtually the whole range of molecules found in living things. Biochemical engineers, then, will construct new enzymes to assemble new patterns of atoms. For example, they might make an enzyme-like machine which will add carbon atoms to a small spot, layer on layer. If bonded correctly, the atoms will build up to form a fine, flexible diamond fiber having over fifty times as much strength as the same weight of aluminum.

Aerospace companies will line up to buy such fibers by the ton to make advanced composites. (This shows one small reason why military competition will drive molecular technology forward, as it has driven so many fields in the past.) But the great advance will come when protein machines are able to make structures more complex than mere fibers. These programmable protein machines will resemble ribosomes programmed by RNA, or the older generation of automated machine tools programmed by punched tapes. They will open a new world of possibilities, letting engineers escape the limitations of proteins to build rugged, compact machines with straightforward designs. Engineered proteins will split and join molecules as enzymes do. Existing proteins bind a variety of smaller molecules, using them as chemical tools; newly engineered proteins will use all these tools and more.

Further, organic chemists have shown that chemical reactions can produce remarkable results even without nanomachines to guide the molecules. Chemists have no direct control over the tumbling motions of molecules in a liquid, and so the molecules are free to react in any way they can, depending on how they bump together. Yet chemists nonetheless coax reacting molecules to form regular structures such as cubic and dodecahedral molecules, and to form unlikely-seeming structures such as molecular rings with highly strained bonds. Molecular machines will have still greater versatility in bond-making, because they can use similar molecular motions to make bonds, but can guide these motions in ways that chemists cannot. Indeed, because chemists cannot yet direct molecular motions, they can seldom assemble complex molecules according to specific plans.

The largest molecules they can make with specific, complex patterns are all linear chains. Chemists form these patterns (as in gene machines) by adding molecules in sequence, one at a time, to a growing chain. With only one possible bonding site per chain, they can be sure to add the next piece in the right place. But if a rounded, lumpy molecule has (say) a hundred hydrogen atoms on its surface, how can chemists split off just one particular atom (the one five up and three across from the bump on the front) to add something in its place? Stirring simple chemicals together will seldom do the job, because small molecules can seldom select specific places to react with a large molecule. But protein machines will be more choosy. A flexible, programmable protein machine will grasp a large molecule (the workpiece) while bringing a small molecule up against it in just the right place. Like an enzyme, it will then bond the molecules together. By bonding molecule after molecule to the workpiece, the machine will assemble a larger and larger structure while keeping complete control of how its atoms are arranged. This is the key ability that chemists have lacked.

Like ribosomes, such nanomachines can work under the direction of molecular tapes. Unlike ribosomes, they will handle a wide variety of small molecules (not just amino acids) and will join them to the workpiece anywhere desired, not just to the end of a chain. Protein machines will thus combine the splitting and joining abilities of enzymes with the programmability of ribosomes. But whereas ribosomes can build only the loose folds of a protein, these protein machines will build small, solid objects of metal, ceramic, or diamond—invisibly small, but rugged. Where our fingers of flesh are likely to bruise or burn, we turn to steel tongs. Where protein machines are likely to crush or disintegrate, we will turn to nanomachines made of tougher stuff.

Universal Assemblers

These second-generation nanomachines built of more than just proteins will do all that proteins can do, and more. In particular, some will serve as improved devices for assembling molecular structures. Able to tolerate acid or vacuum, freezing or baking, depending on design, enzyme-like second-generation machines will be able to use as "tools" almost any of the reactive molecules used by chemists—but they will wield them with the precision of programmed machines. They will be able to bond atoms together in virtually any stable pattern, adding a few at a time to the surface of a workpiece until a complex structure is complete. Think of such nanomachines as assemblers. Because assemblers will let us place atoms in almost any reasonable arrangement, they will let us build almost anything that the laws of nature allow to exist.

In particular, they will let us build almost anything we can design—including more assemblers. The consequences of this will be profound/

because our crude tools have let us explore only a small part of the range of possibilities that natural law permits. Assemblers will open a world of new technologies. Advances in the technologies of medicine, space, computation, and production—and warfare—all depend on our ability to arrange atoms. With assemblers, we will be able to remake our world or destroy it. So at this point it seems wise to step back and look at the prospect as clearly as we can, so we can be sure that assemblers and nanotechnology are not a mere futurological mirage.

Conclusions

In everything we have been describing, we have stuck closely to the demonstrated facts of chemistry and molecular biology. Still, people regularly raise certain questions rooted in physics and biology. These deserve more direct answers.

Will the uncertainly principle of quantum physics make molecular machines unworkable?

This principle states (among other things) that particles can't be pinned down in an exact location for any length of time. It limits what molecular machines can do, just as it limits what anything else can do. Nonetheless, calculations show that the uncertainty principle places few important limits on how well atoms can be held in place, at least for the purposes outlined here. The uncertainty principle makes electron positions quite fuzzy, and in fact this fuzziness determines the very size and structure of atoms. An atom as a whole, however, has a comparatively definite position set by its comparatively massive nucleus. If atoms didn't stay put fairly well, molecules would not exist. One needn't study quantum mechanics to trust these conclusions, because molecular machines in the cell demonstrate that molecular machines work.

Will the molecular vibrations of heat make molecular machines unworkable or too unreliable for use?

Thermal vibrations will cause greater problems than will the uncertainty principle, yet here again existing molecular machines directly demonstrate that molecular machines can work at ordinary temperatures. Despite thermal vibrations, the DNA-copying machinery in some cells makes less than one error in 100,000,000,000 operations. To achieve this accuracy, however, cells use machines (such as the enzyme DNA polymerase 1) that proofread the copy and correct errors. Assemblers may well need similar error-checking and error-correcting abilities, if they are to produce reliable results.

Will radiation disrupt molecular machines and render them unusable?

High-energy radiation can break chemical bonds and disrupt molecular machines. Living cells once again show that solutions exist: they operate for years by repairing and replacing radiation-damaged parts. Because individual machines are so tiny, however, they present small targets for radiation and are seldom hit. Still, if a system of nanomachines must be reliable, then it will have to tolerate a certain amount of damage, and damaged parts must regularly be repaired or replaced. This approach to reliability is well known to designers of aircraft and spacecraft.

Since evolution has failed to produce assemblers, does this show that they are either impossible or useless?

The earlier questions were answered in part by pointing to the working molecular machinery of cells. This makes a simple and powerful case that natural law permits small clusters of atoms to behave as controlled machines, able to build other nanomachines. Yet despite their basic resemblance to ribosomes, assemblers will differ from anything found in cells; the things they do—while consisting of ordinary molecular motions and reactions—will have novel results. No cell, for example, makes diamond fiber.

The idea that new kinds of nanomachinery will bring new, useful abilities may seem startling: in all its billions of years of evolution, life has never abandoned its basic reliance on protein machines. Does this suggest that improvements are impossible, though? Evolution progresses through small changes, and evolution of DNA cannot easily replace DNA. Since the DNA/RNA/ribosome system is specialized to make proteins, life has had no real opportunity to evolve an alternative. Any production manager can well appreciate the reasons; even more than a factory, life cannot afford to shut down to replace its old systems.

Improved molecular machinery should no more surprise us than alloy steel being ten times stronger than bone, or copper wires transmitting signals a million times faster than nerves. Cars outspeed cheetahs, jets outfly falcons, and computers already outcalculate head-scratching humans. The future will bring further examples of improvements on biological evolution, of which second-generation nanomachines will be but one. In physical terms, it is clear enough why advanced assemblers will be able to do more than existing protein machines. They will be programmable like ribosomes, but they will be able to use a wider range of tools than all the enzymes in a cell put together. Because they will be made of materials far more strong, stiff, and stable than proteins, they will be able to exert greater forces, move with greater precision, and endure harsher conditions.

Like an industrial robot arm—but unlike anything in a living cell—they will be able to rotate and move molecules in three dimensions under programmed control, making possible the precise assembly of complex objects. These advantages will enable them to assemble a far wider range of molecular structures than living cells have done.

Is there some special magic about life, essential to making molecular machinery work?

One might doubt that artificial nanomachines could even equal the abilities of nanomachines in the cell, if there were reason to think that cells contained some special magic that makes them work. This idea is called "vitalism." Biologists have abandoned it because they have found chemical and physical explanations for every aspect of living cells yet studied, including their motion, growth, and reproduction. Indeed, this knowledge is the very foundation of biotechnology. Nanomachines floating in sterile test tubes, free of cells, have been made to perform all the basic sorts of activities that they perform inside living cells. Starting with chemicals that can be made from smoggy air, biochemists have built working protein machines without help from cells. R. B. Merrifield, for example, used chemical techniques to assemble simple amino acids to make bovine pancreatic ribonuclease, an enzymatic device that disassembles RNA molecules. Life is special in structure, in behaviour, and in what it feels like from the inside to be alive, yet the laws of nature that govern the machinery of life also govern the rest of the universe.

The case for the feasibility of assemblers and other nanomachines may sound firm, but why not just wait and see whether they can be developed?

Sheer curiosity seems reason enough to examine the possibilities opened by nanotechnology, but there are stronger reasons. These developments will sweep the world within ten to fifty years—that is, within the expected lifetimes of ourselves or our families. It would cost many millions of lives, and perhaps end life on Earth. Is the case for the feasibility of nanotechnology and assemblers firm enough that they should be taken seriously? It seems so, because the heart of the case rests on two well-established facts of science and engineering. These are (1) that existing molecular machines serve a range of basic functions, and (2) that parts serving these basic functions can be combined to build complex machines. Since chemical reactions can bond atoms together in diverse ways, and since molecular machines can direct chemical reactions according to programmed instructions, assemblers definitely are feasible.

Nanocomputers

Assemblers will bring one breakthrough of obvious and basic importance: engineers will use them to shrink the size and cost of computer circuits and speed their operation by enormous factors. With today's bulk technology, engineers make patterns on silicon chips by throwing atoms and photons at them, but the patterns remain flat and molecular-scale flaws are unavoidable. With assemblers, however, engineers will build circuits in three dimensions, and build to atomic precision. The exact limits of electronic technology today remain uncertain because the quantum behaviour of electrons in complex networks of tiny structures presents complex problems, some of them resulting directly from the uncertainty principle. Whatever the limits are, though, they will be reached with the help of assemblers. The fastest computers will use electronic effects, but the smallest may not. This may seem odd, yet the essence of computation has nothing to do with electronics.

A digital computer is a collection of switches able to turn one another on and off. Its switches start in one pattern (perhaps representing 2 + 2), then switch one another into a new pattern (representing 4), and so on. Such patterns can represent almost anything. Engineers build computers from tiny electrical switches connected by wires simply because mechanical switches connected by rods or strings would be big, slow, unreliable, and expensive, today. The idea of a purely mechanical computer is scarcely new. In England during the mid-1800s, Charles Babbage invented a mechanical computer built of brass gears; his co-worker Augusta Ada, the Countess of Lovelace, invented computer programming. Babbage's endless redesigning of the machine, problems with accurate manufacturing, and opposition from budget-watching critics (some doubting the usefulness of computers!), combined to prevent its completion.

In this tradition, Danny Hillis and Brian Silverman of the MIT Artificial Intelligence Laboratory built a special-purpose mechanical computer able to play tic-tactoe. Yards on a side, full of rotating shafts and movable frames that represent the state of the board and the strategy of the game, it now stands in the Computer Museum in

Boston. It looks much like a large ball-and-stick molecular model, for it is built of Tinkertoys.

Brass gears and Tinkertoys make for big, slow computers. With components a few atoms wide, though, a simple mechanical computer would fit within 1/100 of a cubic micron, many billions of times more compact than today's so-called microelectronics. Even with a billion bytes of storage, a nanomechanical computer could fit in a box a micron wide, about the size of a bacterium. And it would be fast. Although mechanical signals move about 100,000 times slower than the electrical signals in today's machines, they will need to travel only 1/1,000,000 as far, and thus will face less delay. So a mere mechanical computer will work faster than the electronic whirl-winds of today.

Electronic nanocomputers will likely be thousands of times faster than electronic microcomputers—perhaps hundreds of thousands of times faster, if a scheme proposed by Nobel Prize-winning physicist Richard Feynman works out. Increased speed through decreased size is an old story in electronics. Disassemblers Molecular computers will control molecular assemblers, providing the swift flow of instructions needed to direct the placement of vast numbers of atoms. Nanocomputers with molecular memory devices will also store data generated by a process that is the opposite of assembly.

Assemblers will help engineers synthesize things; their relatives, disassemblers, will help scientists and engineers analyze things. The case for assemblers rests on the ability of enzymes and chemical reactions to form bonds, and of machines to control the process. The case for disassemblers rests on the ability of enzymes and chemical reactions to break bonds, and of machines to control the process. Enzymes, acids, oxidizers, alkali metals, ions, and reactive groups of atoms called free radicals—all can break bonds and remove groups of atoms. Because nothing is absolutely immune to corrosion, it seems that molecular tools will be able to take anything apart, a few atoms at a time. What is more, a nanomachine could (at need or convenience) apply mechanical force as well, in effect prying groups of atoms free.

A nanomachine able to do this, while recording what it removes layer by layer, is a disassembler. Assemblers, disassemblers, and nanocomputers will work together. For example, a nanocomputer system will be able to direct the disassembly of an object, record its structure, and then direct the assembly of perfect copies. And this gives some hint of the power of nanotechnology.

Assemblers will take years to emerge, but their emergence seems almost inevitable: Though the path to assemblers has many steps, each step will bring the next in reach, and each will bring immediate rewards. The first steps have already been taken, under the names of "genetic engineering" and "biotechnology." Other paths to assemblers seem possible. Barring worldwide destruction or worldwide controls, the technology race will continue whether we wish it or not. And as advances in computer aided design speed the development of molecular tools, the advance toward assemblers will quicken. To have any hope of understanding our future, we must understand the consequences of assemblers, disassemblers, and nanocomputers. They promise to bring changes as profound as the industrial revolution, antibiotics, and nuclear weapons all rolled up in one massive breakthrough. To understand a future of such profound change, it makes sense to seek principles of change that have survived the greatest upheavals of the past. They will prove a useful guide.

10

CELL REPAIR MACHINE

Since the human body is made of molecules, molecular technology will be used to bring health. The ill, the old, and the injured all suffer from misarranged patterns of atoms, whether misarranged by invading viruses, passing time, or swerving cars. Devices able to rearrange atoms will be able to set them right. *Nanotechnology* will bring a fundamental breakthrough in medicine. Physicians now rely chiefly on surgery and drugs to treat illness. Surgeons have advanced from stitching wounds and amputating limbs to repairing hearts and reattaching limbs. Using microscopes and fine tools, they join delicate blood vessels and nerves. Yet even the best microsurgeon cannot cut and stitch finer tissue structures. Modern scalpels and sutures are simply too coarse for repairing capillaries, cells, and molecules. Consider "delicate" surgery from a cell's perspective a huge blade sweeps down, chopping blindly past and through the molecular machinery of a crowd of cells, slaughtering thousands. Later, a great obelisk plunges through the divided crowd, dragging a cable as wide as a freight train behind it to rope the crowd together again. From a cell's perspective even the most delicate surgery, performed with exquisite knives and great skill, is still a butcher job. Only the ability of cells to abandon their dead, regroup, and multiply makes healing possible. Yet as many paralyzed accident victims know too well, not all tissues heal.

Drug therapy, unlike surgery, deals with the finest structures in cells. *Drug molecules* are simple molecular devices. Many affect specific molecules in cells. *Morphine molecules*, for example, bind to certain receptor molecules in brain cells, affecting the neural impulses that signal pain. *Insulin*, *beta blockers*, and other drugs fit other receptors. But drug molecules work without direction. Once dumped into the body, they tumble and bump around in solution haphazardly until they bump a target molecule, fit, and stick, affecting its function. Surgeons can see problems and plan actions, but they wield crude tools; drug molecules affect tissues at the molecular level, but they are too simple to sense, plan, and act. But molecular machines directed by nano computers will offer physicians another choice. They will combine sensors, programs, and molecular tools to form systems able to examine and repair the ultimate components of individual cells. They will bring surgical control to the molecular domain.

These advanced molecular devices will be years in arriving, but researchers motivated by medical needs are already studying molecular machines and molecular engineering. The best drugs affect specific molecular machines in specific ways. Penicillin, for example, kills certain bacteria by jamming the *nanomachinery* they use to build their cell walls, yet it has little effect on human cells.

Biochemists study molecular machines both to learn how to build them and to learn how to wreck them. Around the world (and especially the Third World) a disgusting variety of viruses, bacteria, protozoa, fungi, and worms parasitize human flesh. Like penicillin, safe, effective drugs for these diseases would jam the parasite's molecular machinery while leaving human molecular machinery unharmed. Dr. Seymour Cohen, professor of pharmacological science at Suny, argues that biochemists should systematically study the molecular machinery of these parasites. Once biochemists have determined the shape and function of a vital protein machine, they then could often design a molecule shaped to jam it and ruin it. Such drugs could free humanity from such ancient horrors as schistosomiasis and leprosy, and from new ones such as AIDS.

Drug companies are already redesigning molecules based on knowledge of how they work. Researchers at Upjohn Company have designed and made modified molecules of *vasopressin*, a hormone that consists of a short chain of amino acids. Vasopressin increases the work done by the heart and decreases the rate at which the kidneys produce urine; this increases blood pressure. The researchers de signed modified vasopressin molecules that affected receptor molecules in the kidney more than those in the heart, giving them more specific and controllable medical effects. More recently, they de signed a modified vasopressin molecule that binds to the kidney's receptor molecules without direct effect, thus blocking and *inhibiting* the action of natural vasopressin.

Medical needs will push this work forward, encouraging researchers to take further steps toward protein design and molecular engineering. Medical, military, and economic pressures all push us in the same direction. Even before the assembler breakthrough, *molecular technology* will bring impressive advances in medicine; trends in biotechnology guarantee it. Still, these advances will generally be piecemeal and hard to predict, each exploiting some detail of biochemistry. Later, when we apply assemblers and technical AI systems to medicine, we will gain broader abilities that are easier to foresee.

To understand these abilities, consider cells and their self-repair mechanisms. In the cells of your body, natural radiation and noxious chemicals split molecules, producing reactive molecular fragments. These can misbond to other molecules in a process called cross-linking. As bullets and blobs of glue would damage a machine, so radiation and reactive fragments damage cells, both breaking molecular machines and gumming them up. If your cells could not repair themselves, damage would rapidly kill them or make them run amok by damaging their control systems. But evolution has favoured organisms with machinery able to do something about this problem. The self-replicating factory system repaired itself by replacing damaged parts; cells do the same. So long as a cell's DNA remains intact, it can make error-free tapes that direct ribosomes to assemble new protein machines.

Unfortunately for us, DNA itself becomes damaged, resulting in mutations. Repair enzymes compensate somewhat by detecting and repairing certain kinds of damage to DNA. These repairs help cells survive, but existing repair mechanisms are too simple to correct all problems, either in DNA or elsewhere. Errors mount, contributing to the aging and death of cells—and of people.

Life, Mind, and Machines

Does it make sense to describe cells as "machinery," whether self- repairing or not? Since we are made of cells, this might seem to reduce human beings to "*mere machines*," conflicting with a holistic understanding of life.

But a dictionary definition of holism is "the theory that reality is made up of organic or unified wholes that are greater than the simple sum of their parts." This certainly applies to people: one simpler sum of our parts would resemble hamburger, lacking both mind and life. The human body includes some ten thousand billion *billion* protein parts, and no machine so complex deserves the label "mere." Any brief description of so complex a system cannot avoid being grossly incomplete, yet at the cellular level a description in terms of machinery makes sense. Molecules have simple moving parts, and many act like familiar types of machinery. Cells considered as a whole may seem less mechanical, yet biologists find it useful to describe them in terms of molecular machinery.

Biochemists have unraveled what were once the central mysteries of life, and have begun to fill in the details. They have traced how molecular machines break food molecules into their building blocks and then reassemble these parts to build and renew tissue. Many details of the structure of human cells remain unknown (single cells have billions of large molecules of thousands of different kinds), but biochemists have mapped every part of some viruses. Biochemical laboratories often sport a large wall chart showing how the chief molecular building blocks flow through bacteria. Biochemists under stand much of the process of life in detail, and what they don't understand seems to operate on the same principles. The mystery of heredity has become the industry of genetic engineering. Even embryonic development and memory are being explained in terms of changes in biochemistry and cell structure.

In recent decades, the very quality of our remaining ignorance has changed. Once, biologists looked at the process of life and asked, "How can this be?" But today they understand the general principles of life, and when they study a specific living process they commonly ask, "Of the many ways this could be, which has nature chosen?" In many instances their studies have narrowed the competing explanations to a field of one. Certain biological processes—the coordination of cells to form growing embryos, learning brains, and reacting immune systems-still present a real challenge to the imagination. Yet this is not because of some deep mystery about how their parts work, but because of the immense complexity of how their many parts interact to form a whole.

Cells obey the same natural laws that describe the rest of the world. *Protein machines* in the right molecular environment will work whether they remain in a functioning cell or whether the rest of the cell was ground up and washed away days before. Molecular machines know nothing of "*life*" and "*death*."

Biologists—when they bother—sometimes define life as the ability to grow, replicate, and respond to stimuli. But by this standard, a mindless system of replicating factories might qualify as life, while a conscious artificial intelligence modeled on the human brain might not. Are viruses alive, or are they "*merely*" fancy molecular machines? No experiment can tell, because nature draws no line between living and nonliving. Biologists who work with viruses instead ask about viability: "Will this virus function, if given a chance?" The labels of "*life*" and "*death*" in medicine depend on medical capabilities: physicians ask, "Will this patient function, if we do our best?" Physicians once declared patients dead when the heart stopped; they now declare patients dead when they despair of restoring brain activity. Advances in cardiac medicine changed the definition once; advances in brain medicine will change it again.

Just as some people feel uncomfortable with the idea of machines thinking, so some feel uncomfortable with the idea that machines underlie our own thinking. The word "*machine*" again seems to conjure up the wrong image, a picture of gross, clanking metal, rather than signals flickering through a shifting weave of neural fibers, through a living tapestry more intricate than the mind it embodies can fully comprehend. The brain's really machinelike machines are of molecular size, smaller than the finest fibers.

A whole need not resemble its parts. A solid lump scarcely resembles a dancing fountain, yet a collection of solid, lumpy molecules forms fluid water. In a similar way, billions of molecular machines make up neural fibers and synapses, thousands of fibers and synapses make up a neural cell, billions of neural cells make up the brain, and the brain itself embodies the fluidity of thought. To say that the mind is "*just molecular machines*" is like saying that the Mona Lisa is '*just dabs of paint.*" Such statements confuse the parts with the whole, and confuse matter with the pattern it embodies. We are no less human for being made of molecules.

From Drugs to Cell Repair Machines

Being made of molecules, and having a human concern for our health, we will apply molecular machines to biomedical technology. Biologists already use antibodies to tag proteins, enzymes to cut and splice DNA, and viral syringes (like the T_4 phage) to inject edited DNA into bacteria. In the future, they will use assembler-built nanomachines to probe and modify cells. With tools like disassemblers, biologists will be able to study cell structures in ultimate, molecular detail. They then will catalog the hundreds of thousands of kinds of molecules in the body and map the structure of the hundreds of kinds

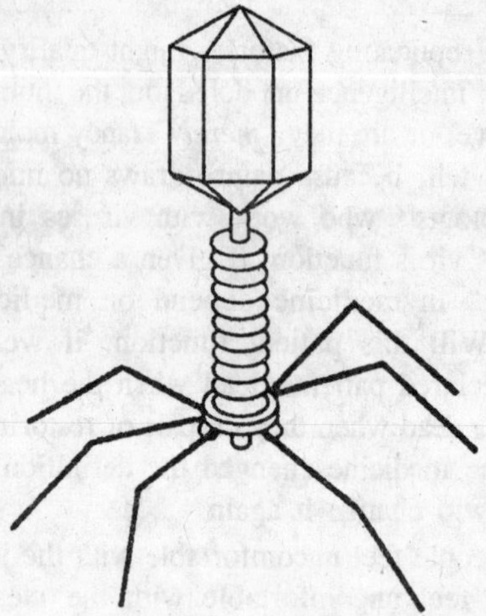

Fig. 10.1. Bacteriophage T_4

of cells. Much as engineers might compile a parts list and make engineering drawings for an automobile, so biologists will describe the parts and structures of healthy tissue. By that time, they will be aided by sophisticated technical AI systems. Physicians aim to make tissues healthy, but with drugs and surgery they can only encourage tissues to repair themselves. Molecular machines will allow more direct repairs, bringing a new era in medicine.

To repair a car, a mechanic first reaches the faulty assembly, then identifies and removes the bad parts, and finally rebuilds or replaces them. Cell repair will involve the same basic tasks–tasks that living systems already prove possible.

Access

White blood cells leave the bloodstream and move through tissue, and viruses enter cells. Biologists even poke needles into cells without killing them. These examples show that molecular machines can reach and enter cells.

Recognition

Antibodies and the tail fibers of the T_4 phage–and indeed, all specific biochemical interactions–show that molecular systems can recognize other molecules by touch.

Disassembly

Digestive enzymes (and other, fiercer chemicals) show that molecular systems can disassemble damaged molecules.

Rebuilding

Replicating cells show that molecular systems can build or rebuild every molecule found in a cell.

Reassembly

Nature also shows that separated molecules can be put back together again. The machinery of the T_4 phage, for example, self-assembles from solution, apparently aided by a single enzyme. Replicating cells show that molecular systems can assemble every system found in a cell.

Thus, nature demonstrates all the basic operations that are needed to perform molecular-level repairs on cells. What is more, systems based on nanomachines will generally be more compact and capable than those found in nature. Natural systems show us only lower bounds to the possible, in cell repair as in everything else.

Cell Repair Machines

In short, with molecular technology and technical AI we will compile complete, molecular-level descriptions of healthy tissue, and we will build machines able to enter cells and to sense and modify their structures.

Cell repair machines will be comparable in size to bacteria and viruses, but their more-compact parts will allow them to be more complex. They will travel through tissue as white blood cells do, and enter cells as viruses do—or they could open and close cell membranes with a surgeon's care. Inside a cell, a repair machine will first size up the situation by examining the cell's contents and activity, and then take action. Early cell repair machines will be highly specialized, able to recognize and correct only a single type of molecular disorder, such as an enzyme deficiency or a form of DNA damage. Later machines (but not much later, with advanced technical AI systems doing the design work) will be programmed with more general abilities.

Complex repair machines will need nanocomputers to guide them. A micron-wide mechanical computer will fit in 1/1000 of the volume of a typical cell, yet will hold more information than does the cell's DNA. In a repair system, such computers will direct smaller, simpler computers, which will in turn direct machines to examine, take apart, and rebuild damaged molecular structures.

By working along molecule by molecule and structure by structure, repair machines will be able to repair whole cells. By working along cell by cell and tissue by tissue, they (aided by larger devices, where need be) will be able to repair whole organs. By working through a person organ by organ, they will restore health. Because molecular machines will be able to build molecules and cells from scratch, they will be able to repair even cells damaged to the point of complete inactivity. Thus, cell repair machines will bring a fundamental breakthrough: they will free medicine from reliance on self- repair as the only path to healing.

To visualize an advanced cell repair machine, imagine it—and a cell—enlarged until atoms are the size of small marbles. On this scale, the repair machine's smallest tools have tips about the size of your fingertips; a medium-sized protein, like hemoglobin, is the size of a typewriter; and a ribosome is the size of a washing machine. A single repair device contains a simple computer the size of a small truck, along with many sensors of protein size, several manipulators of ribosome size, and provisions for memory and motive power. A total volume ten meters across, the size of a three-story house, holds all these parts and more. With parts the size of marbles packing this volume, the repair machine can do complex things.

But this repair device does not work alone. It, like its many siblings, is connected to a larger computer by means of mechanical data links the diameter of your arm. On this scale, a cubic-micron computer with a large memory fills a volume thirty stories high and as wide as a football field. The repair devices pass it information, and it passes back general instructions. Objects so large and complex are still small enough: on this scale, the cell itself is a kilometer across, holding one thousand times the volume of a cubic-micron computer, or a million times the volume of a single repair device. Cells are spacious.

Will such machines be able to do everything necessary to repair cells? Existing molecular machines demonstrate the ability to travel through tissue, enter cells, recognize molecular structures, and so forth, but other requirements are also important. Will repair machines work fast enough? If they do, will they waste so much power that the patient will roast?

The most extensive repairs cannot require *vastly* more work than building a cell from scratch. Yet molecular machinery working within a cellular volume routinely does just that, building a new cell in tens of minutes (in bacteria) to a few hours (in mammals). This indicates that repair machinery occupying a few percent of a cell's volume will be able to complete even extensive repairs in a reason able time—

days or weeks at most. Cells can spare this much room. Even brain cells can still function when an inert waste called *lipofuscin* (apparently a product of molecular damage) fills over ten percent of their volume.

Powering repair devices will be easy: cells naturally contain chemicals that power *nanomachinery*. Nature also shows that repair machines can be cooled: the cells in your body rework themselves steadily, and young animals grow swiftly without cooking themselves. Handling heat from a similar level of activity by repair machines will be no sweat—or at least not too much sweat, if a week of sweating is the price of health.

All these comparisons of repair machines to existing biological mechanisms raise the question of whether repair machines will be able to *improve* on nature. DNA repair provides a clear-cut illustration.

Just as an illiterate "*book-repair machine*" could recognize and repair a torn page, so a cell's repair enzymes can recognize and repair breaks and cross-links in DNA. Correcting misspellings (or mutations), though, would require an ability to read. Nature lacks such repair machines, but they will be easy to build. Imagine three identical DNA molecules, each with the same sequence of nucleotides. Now imagine each strand mutated to change a few scattered nucleotides. Each strand still seems normal, taken by itself. Nonetheless, a repair machine could compare each strand to the others, one segment at a time, and could note when a nucleotide failed to match its mates. Changing the odd nucleotide to match the other two will then repair the damage. This method will fail if two strands mutate in the same spot. Imagine that the DNA of three human cells has been heavily damaged— after thousands of mutations, each cell has had one in every million nucleotides changed. The chance of our three-strand correction procedure failing at any given spot is then about one in a million mil lion. But compare five strands at once, and the odds become about one in a million million-million, and so on. A device that compares many strands will make the chance of an uncorrectable error effectively nil.

In practice, repair machines will compare DNA molecules from several cells, make corrected copies, and use these as standards for proofreading and repairing DNA throughout a tissue. By comparing several strands, repair machines will dramatically improve on nature's repair enzymes.

Other repairs will require different information about healthy cells and about how a particular damaged cell differs from the norm. Anti bodies identify proteins by touch, and properly chosen antibodies can generally distinguish any two proteins by their differing shapes and surface properties. Repair machines will identify molecules in a similar way. With a suitable computer and data base, they will be able to identify proteins by reading their amino acid sequences.

Consider a complex and capable repair system. A volume of two cubic microns—about 2/1000 of the volume of a typical cell—will be enough to hold a central data base system able to:

1. Swiftly identify any of the hundred thousand or so different human proteins by examining a short amino acid sequence.
2. Identify all the other complex molecules normally found in cells.
3. Record the type and position of every large molecule in the cell.

Each of the smaller repair devices (of perhaps thousands in a cell) will include a less capable computer. Each of these computers will be able to perform over a thousand computational steps in the time that a typical enzyme takes to change a single molecular bond, so the speed of computation possible seems more than adequate. Because each computer will be in communication with a larger computer and the central data base, the available memory seems adequate. Cell repair machines will have both the molecular tools they need and "brains" enough to decide how to use them.

Such sophistication will be overkill (over cure?) for many health problems. Devices that merely recognize and destroy a specific kind of cell, for example, will be enough to cure a cancer. Placing a computer network in every cell may seem like slicing butter with a chain saw, but having a chain saw available does provide assurance that even hard butter can be sliced. It seems better to show too much than too little, if one aims to describe the limits of the possible in medicine.

Some Cures

The simplest medical applications of *nanomachines* will involve not repair but selective destruction. Cancers provide one example; infectious diseases provide another. The goal is simple: one need only recognize and destroy the dangerous replicators, whether they are bacteria, cancer cells, viruses, or worms. Similarly, abnormal growths and deposits on arterial walls cause much heart disease; machines that recognize, break down, and dispose of them will clear arteries for more normal blood flow. Selective destruction will also cure diseases such as herpes in which a virus splices its genes into the DNA of a host cell. A repair device will enter the cell, read its DNA, and remove the addition that spells "*herpes.*"

Repairing damaged, cross-linked molecules will also be fairly straightforward. Faced with a damaged, cross-linked protein, a cell repair machine will first identify it by examining short amino acid sequences, then look up its correct structure in a data base. The machine will then compare the protein to this blueprint, one amino acid at a time. Like a proofreader finding misspellings and strange characters (characters), it will find any changed amino acids or improper cross-links. By correcting these flaws, it will leave a normal protein, ready to do the work of the cell.

Repair machines will also aid healing. After a heart attack, scar tissue replaces dead muscle. Repair machines will stimulate the heart to grow fresh muscle by resetting cellular control mechanisms. By removing scar tissue and guiding fresh growth, they will direct the healing of the heart.

This list could continue through problem after problem (*Heavy metal poisoning*? —Find and remove the metal atoms) but the conclusion is easy to summarize. Physical disorders stem from misarranged atoms; repair machines will be able to return them to working order, restoring the body to health, Rather than compiling an endless list of curable diseases (from *arthritis*, *bursitis*, cancer, and dengue to yellow fever and zinc chills and back again), it makes sense to look for the limits to what cell repair machines can do. Limits do exist.

Consider stroke, as one example of a problem that damages the brain. Prevention will be straightforward: Is a blood vessel in the brain weakening, bulging, and apt to burst? Then pull it back into shape and guide the growth of reinforcing fibers. Does abnormal cloning threaten to block circulation? Then dissolve the clots and normalize the blood and blood-vessel linings to prevent a recurrence. Moderate neural damage from stroke will also be repairable: if reduced circulation has impaired function but left cell structures intact, then restore circulation and repair the cells, using their structures as a guide in restoring the tissue to its previous state. This will not only restore each cell's function, but will preserve the memories and skills embodied in the neural patterns in that part of the brain.

Repair machines will be able to regenerate fresh brain tissue even where damage has obliterated these patterns. But the patient would lose old memories and skills to the extent that they resided in that pan of the brain. If unique neural patterns are truly obliterated, then cell repair machines could no more restore them than art conservators could restore a tapestry from stirred ash. Loss of information through obliteration of structure imposes the most important, fundamental limit to the repair of tissue.

Other tasks are beyond cell repair machines for different reasons—maintaining mental health, for instance. Cell repair machines will be able to correct some problems, of course. Deranged thinking some times has biochemical causes, as if the brain were drugging or poisoning itself, and other problems stem from tissue damage. But many problems have little to do with the health of nerve cells and everything to do with the health of the mind.

A mind and the tissue of its brain are like a novel and the paper of its book. Spilled ink or flood damage may harm the book, making the novel difficult to read. Book repair machines could nonetheless restore physical "*health*" by removing the foreign ink or by drying and repairing the damaged paper fibers. Such treatments would do nothing for the book's *content*, however, which in a real sense is nonphysical. If the book were a cheap romance with a moldy plot and empty characters, repairs would be needed not on the ink and paper, but on the novel. This would call not for physical repairs, but for more work by the author, perhaps with advice.

Similarly, removing poisons from the brain and repairing its nerve fibers will thin some mental fogs, but not revise the content of the mind. This can be changed by the patient, with effort; we are all authors of our minds, But because minds change themselves by changing their brains, having a healthy brain will aid sound thinking more than quality paper aids sound writing. Readers familiar with computers may prefer to think in terms of hardware and software. A machine could repair a computer's hard ware while neither understanding nor changing its software.

Such machines might stop the computer's activity but leave the patterns in memory intact and ready to work again. In computers with the right kind of memory (called *nonvolatile*), users do this by simply switching off the power. In the brain the job seems more complex, yet there could be medical advantages to inducing a similar state.

Anesthesia Plus

Physicians already stop and restart consciousness by interfering with the chemical activity that underlies the mind. Throughout active life, molecular machines in the brain process molecules. Some disassemble sugars, combine them with oxygen, and capture the energy this releases. Some pump salt ions across cell membranes; others build small molecules and release them to signal other cells. Such processes make up the brain's metabolism, the sum total of its chemical activity. Together with its electrical effects, this metabolic activity underlies the changing patterns of thought.

Surgeons cut people with knives. In the mid-1800s, they learned to use chemicals that interfere with brain metabolism, blocking conscious thought and preventing patients from objecting so vigorously to being cut. These chemicals are *anesthetics*. Their molecules freely enter and leave the brain, allowing anesthetists to interrupt and restart human consciousness.

People have long dreamed of discovering a drug that interferes with the metabolism of the entire body, a drug able to interrupt metabolism completely for hours, days, or years. The result would be a condition of *biostasis* (from *bio*, meaning life, and *stasis*, meaning a stoppage or a stable state). A method of producing *reversible biostasis* could help *astronauts* on long space voyages to save food and avoid boredom, or it could serve as a kind of one-way time travel. In medicine, *biostasis* would provide a deep *anesthesia* giving physicians more time to work. When emergencies occur far from medical help, a good biostasis procedure would provide a sort of universal first-aid treatment: it would stabilize a patient's condition and prevent molecular machines from running amok and damaging tissues.

But no one has found a drug able to stop the entire metabolism the way anesthetics stop consciousness—that is, in a way that can be reversed by simply washing the drug out of the patient's tissues. Nonetheless, *reversible biostasis* will be possible when repair machines become available.

To see how one approach would work, imagine that the blood stream carries simple molecular devices to tissues, where they enter the cells. There they block the molecular machinery of metabolism—in the brain and elsewhere—and tie structures together with stabilizing cross-links. Other molecular devices then move in, displacing water and packing themselves solidly around the molecules of the cell. These steps stop metabolism and preserve cell structures. Be cause cell repair machines will be used to reverse this process, it can cause moderate molecular damage and yet do no lasting harm. With metabolism stopped and cell structures held firmly in place, the patient will rest quietly, dreamless and unchanging, until repair machines restore active life.

If a patient in this condition were turned over to a present-day physician ignorant of the capabilities of cell repair machines, the consequences would likely be grim. Seeing no signs of life, the physician would likely conclude that the patient was dead, and then would make this judgment a reality by "*prescribing*" an *autopsy*, followed by burial or *burning*.

But our imaginary patient lives in an era when biostasis is known to be only an interruption of life, not an end to it. When the patient's contract says "*wake me*!" (or the repairs are complete, or the flight to the stars is finished), the attending physician begins resuscitation. Repair machines enter the patient's tissues, removing the packing from around the patient's molecules and replacing it with water. They then remove the cross-links, repair any damaged molecules and structures, and restore normal concentrations of salts, blood sugar, ATP, and so forth. Finally, they unblock the metabolic machinery. The interrupted metabolic processes resume, the patient yawns, stretches, sits up, thanks the doctor, checks the date, and walks out the door.

From Function to Structure

The reversibility of biostasis and irreversibility of severe stroke damage help to show how cell repair machines will change medicine. Today, physicians can only help tissues to heal themselves. Accordingly, they must try to preserve the *function* of tissue. If tissues cannot function, they cannot heal. Worse, unless they are preserved, deterioration follows, ultimately obliterating structure. It is as if a mechanic's tools were able to work only on a running engine.

Cell repair machines change the central requirement from preserving *function* to preserving *structure*. Repair machines will be able to restore brain function with memory and skills intact only if the distinctive structure of the neural fabric remains intact. Biostasis involves preserving neural structure while deliberately blocking function. All this is a direct consequence of the molecular nature of the repairs. Physicians using scalpels and drugs can no more repair cells than someone using only a pickax and a can of oil can repair a fine watch. In contrast, having repair machines and ordinary nutrients will be like having a watchmaker's tools and an unlimited supply of spare parts. Cell repair machines will change medicine at its foundations.

From Treating Disease to Establishing Health

Medical researchers now study diseases, often seeking ways to pre vent or reverse them by blocking a key step in the disease process.

The resulting knowledge has helped physicians greatly: they now prescribe insulin to compensate for diabetes, antihypertensive to prevent stroke, penicillin to cure infections, and so on down an impressive list. Molecular machines will aid the study of diseases, yet they will make understanding disease far less important. Repair machines will make it more important to understand health.

The body can be ill in more ways than it can be healthy. Healthy muscle tissue, for example, varies in relatively few ways: it can be stronger or weaker, faster or slower, have this antigen or that one, and so forth. Damaged muscle tissue can vary in all these ways, yet also suffer from any combination of strains, tears, viral infections, parasitic worms, bruises, punctures, poisons, sarcomas, wasting diseases, and congenital abnormalities. Similarly, though neurons are woven in as many patterns as there are human brains, individual synapses and dendrites come in a modest range of forms—if they are healthy.

Once biologists have described normal molecules, cells, and tissues, properly programmed repair machines will be able to cure even unknown diseases. Once researchers describe the range of structures that (for example) a healthy liver may have, repair machines exploring a malfunctioning liver need only look for differences and correct them. Machines ignorant of a new poison and its effects will still recognize it as foreign and remove it. Instead of fighting a million strange diseases, advanced repair machines will establish a state of health.

Developing and programming cell repair machines will require great effort, knowledge, and skill. Repair machines with broad capabilities seem easier to build than to program. Their programs must contain detailed knowledge of the hundreds of kinds of cells and the hundreds of thousands of kinds of molecules in the human body. They must be able to map damaged cellular structures and decide how to correct them. How long will such machines and programs take to be developed? Offhand, the state of biochemistry and its present rate of advance might suggest that the basic knowledge alone will take centuries to collect. But we must beware of the illusion that advances will arrive in isolation.

Repair machines will sweep in with a wave of other technologies. The assemblers that build them will first be used to build instruments for analyzing cell structures. Even a pessimist might agree that human biologists and engineers equipped with these tools could build and program advanced cell repair machines in a hundred years of steady work. A cocksure, far-seeing pessimist might say a thousand years. A really committed nay-sayer might declare that the job would take people a million years. Very well: fast technical AI systems—a million fold faster than scientists and engineers—will then develop advanced cell repair machines in a single calendar year.

A Disease Called "Aging"

Aging is natural, but so were smallpox and our efforts to prevent it. We have conquered smallpox, and it seems that we will conquer aging. *Longevity* has increased during the last century, but chiefly be cause better sanitation and drugs have reduced bacterial illness. The basic human lifespan has increased little.

Still, researchers have made progress toward understanding and slowing the aging process. They have identified some of its causes, such as uncontrolled cross-linking. They have devised partial treatments, such as *antioxidants* and *free-radical inhibitors*. They have proposed and studied other mechanisms of aging, such as "*clocks*" in the cell and changes in the body's hormone balance. In laboratory experiments, special drugs and diets have extended the lifespan of mice by 25 to 45 percent.

Such work will continue; as the baby boom generation ages, expect a boom in aging research. One biotechnology company, Senetek of Denmark, specializes in *aging research*. In April 1985, Eastman Kodak and ICN Pharmaceuticals were reported to have joined in a $45 million venture to produce *isoprinosine* and other drugs with the potential to extend lifespan. The results of conventional antiaging research may substantially lengthen human life spans and improve the health of the old—during the next ten to twenty years. How greatly will drugs, surgery, exercise, and diet extend lifespans? For now, estimates must remain guesswork. Only new scientific knowledge can rescue such predictions from the realm of speculation, because they rely on new science and not just new engineering.

With cell repair machines, however, the potential for life extension becomes clear. They will be able to repair cells so long as their distinctive structures remain intact, and will be able to replace cells that have been destroyed. Either way, they will restore health. Aging is fundamentally no different from any other physical disorder; it is no magical effect of calendar dates on a mysterious life-force. Brittle bones, wrinkled skin, low enzyme activities, slow wound healing, poor memory, and the rest all result from damaged molecular machinery, chemical imbalances, and misarranged structures. By restoring all the cells and tissues of the body to a youthful structure, repair machines will restore youthful health.

People who survive intact until the time of cell repair machines will have the opportunity to regain youthful health and to keep it almost as long as they please. Nothing can make a person (or any thing else) last forever, of course, but barring severe accidents, those wishing to do so will live for a long, long time. As a technology develops, there comes a time when its principles become clear, and with them many of its consequences. The principles of rocketry were clear in the 1930s, and with them the consequence of spaceflight. Filling in the details involved designing and testing tanks, engines, instruments, and so forth. By the early 1950s, many details were known. The ancient dream of flying to the Moon had became a goal one could plan for.

The principles of molecular machinery are already clear, and with them the consequence of cell repair machines. Filling in the details will involve designing molecular tools, assemblers, computers, and so forth, but many details of existing molecular machines are known today. The ancient dream of achieving health and long life has be come a goal one can plan for.

Medical research is leading us, step by step, along a path toward molecular machinery. The global competition to make better materials, electronics, and biochemical tools is pushing us in the same direction. Cell repair machines will take years to develop, but they lie straight ahead. They will bring many abilities, both for good and for ill. A moment's thought about military *replicators* with abilities like those of cell repair machines is enough to turn up nauseating possibilities.

11

INTERPHASIC CHIP

The responses of biological systems to drugs or toxins are typically measured at supracellular levels, ranging from tissue cultures to whole organisms. Some measurements have become possible at the level of single cells by the use of fluorescent indicators (e.g., fluorescent protein indicators of gene expression, calcium concentration, or receptor binding and internalization). A number of assay systems of these kinds has been commercialized. The further development of cellular scale detection, and the development of models of cellular scale responses, would contribute greatly to the discovery of new pharmaceuticals and protective agents.

Some issues relating to the development of cellular scale detection methods and models include the following:

1. Some biopolymers are present in small numbers in the cell, so that noise due to small sample sizes should be considered.
2. Electronic detection of bioactive molecules by nanoscale systems, such as carbon nanotubes, is currently being explored.
3. Optical nanosensors for use in single cells are also currently being explored.
4. What other ideas might be worth considering? Inverted receptors, with normal cytoplasmic readout side available outside the cell for signal transduction? Functionalized viral capsids or other structures that might be endocytosed?

Problem

Imagine being able to take measurements on a single cell, to get accurate readouts that overcome the immense challenges of low signal outputs and interference, with a precision and an effectiveness that allows for early detection of disease and readily gauges the effectiveness of new drug therapies.

The concept of extracting information from cells is not new. For decades, electrophysiologists have prodded and poked at cells, extracting limited amounts of data on physiological cell states and processes. But we felt that the envelope could be pushed further, that a "nanoprobe" could be created to garner virtually unlimited data from a single cell.

Solution

The first task: defining the problem. Microarray technology, which essentially places a human on a chip, was initially viewed as key to building a cell-chip interface, as were standard techniques for measuring protein concentrations and other physiological states inside a cell.

Eventually, the question came back to nanotechnology and how it could be used to find a new solution to an old problem. What more could nanotechnology offer that existing technologies did not already offer?

Would interrogating individual cells be possible? Could it be done *in vivo*? Would it be more useful to focus first on an *in vitro* model? Given federal laws governing scientific research on human beings, we realized that an *in vitro* model would be a more practical approach, at least initially.

But even working with cells in a laboratory comes with its own set of problems. The finicky nature of human cells, along with the challenge of artificially inducing sickness or disease, led some to question the necessity of using human cells at all.

"If we can build up an artificial cell, then we won't have any problems delivering nanoprobes," said Dan Luo, an assistant professor of biological and environmental engineering at Cornell University.

Edward Sargent, an associate professor of electrical and computer engineering at the University of Toronto, liked the idea, pointing out that one could remove the contents of a cell, then add organelles back one at a time for greater control over an experiment.

But Barbara Baird, director at Cornell University's Nanobiotechnology Center, pointed out that artificial cells are very different from real cells. "I don't think anyone would believe an artificial cell would be a good model," Baird said.

The focus of the discussion gradually began to shift to the next major task: conceptualizing the nanoprobe and its potential—and potential limitations. All agreed that the nanoprobe should act like a camera, taking "snapshots" of the cell at given points in time. Yet what would the nanoprobe measure and report back?

But scientists do not know expression profiles of all diseases, especially at the intracellular level. Compounding the problem, as some members pointed out, would be the challenge of designing a probe specific enough to detect only the expression profiles of interest.

Others pointed to the difficulties of interpreting the profiles, and that was assuming such data could be obtained in the first place. The group agreed that a baseline must first be established to gauge changes to a cell in response to single or multiple challenges. However, given cell-to-cell variability, the group acknowledged the enormous challenge of developing accurate baseline readings for a single cell.

Also brought up was the question of scale—that is, how big the nanoprobe would be. The group agreed that the nanoprobe should package the maximum number of parameters into a minimum amount of space. However, figuring out how many parameters could fit into a single probe was not something group members could resolve until they knew what the parameters were going to be.

Then there was the practical question of what the probe would be made of. Would it be encapsulated? If so, how? Resolving these manufacturing questions, the group realized, was not feasible in the time allotted.

Some group members, for the sake of discussion, suggested 10 nanometers as the theoretical diameter of the nanoprobe and asked the group whether a cell could take it up. Their colleagues had the answer: with an average diameter of 10 micrometers, a cell easily could take up a 10-nanometer probe With this discussion came the question of whether the probe should enter the cell like a submarine or take all of its readings from the outer surface of the cell membrane. Some group members also asked whether it would be necessary to get the probe back out. Everyone agreed that keeping the probe outside the cell would be easier and pose less of a problem, but the group also realized that far more data could be gathered if the probe also entered the cell.

The answer, then, at least from a theoretical basis, was simple—the probe would almost certainly have to enter the cell. But would it be just one probe? The group discussed the possibility of a number of different probes, some inside the cell and others outside. These probes, supplemented by other cell-targeted techniques, such as dielectric constant spectroscopy, stretching, and compressing, were all viewed as potentially useful in extracting the maximum amount of data from a single cell.

The next logical concern was whether the nanoprobe might have perturbations on normal cellular processes and affect the readouts. A nanoprobe measuring 10 nanometers in diameter would only take up about a billionth of the volume of the average mammalian cell, but because cells have no free space inside, group members pointed out that any volume added to the cytoplasm could have detrimental effects. However, others pointed to the inevitability of invasiveness and advised against dwelling on it.

Also discussed was a viable method of signal detection, given the small size of the nanoprobe and its consequently small signals. Group members pointed out the pros and cons of chemical, electrical, magnetic, and fluorescent signals. They debated using different combinations of these signals and discussed how best to amplify the signals.

Some suggested that detecting tiny signals would be accomplished most effectively by using fluorescent dyes instead of using such instruments as mechanotransducers and magnetic detectors. Others pointed to the efficacy of indicators that track everything from DNA methylation states and gene mutation to cell cycle checkpoints and viscosity.

Throughout the focus group discussions, each issue discussed seemed to raise an entirely different set of questions and challenges. During the first day of talks, group members realized that, because of the size of the group, the discussion was unfocused and not everyone was on the same page or necessarily putting priorities and objectives in the same place. But the group devised an effective solution: each member prioritized his or her goals for the discussion by writing on a sticky note the biggest problem or objective of building the cell-chip interface.

The sticky notes were placed on a white-board and grouped into categories. Several members saw the delivery of the nanoprobes as the key issue to resolve. Others saw signal detection as the biggest hurdle. Still others saw resolution, readout, and controlling the cell's activity as key issues to discuss. Once these priorities had been established, more focused discussions were possible.

By the end of the conference, group members proudly reported that they had developed an outline for interrogating an individual cell using nanotechnology. The theoretical protocol was as follows:

1. Position a cell over the nanoprobes.
2. Use mechanical force to impale the cell on a microfluidic array.
3. Create a cellular activity profile using nanoprobes that act as mechanical and chemical sensors.
4. Compare the activity profiles of healthy and diseased cells, to identify diseased states of individual cells.

While the focus group did not generate a detailed plan of attack—which was not the goal of the conference—group members learned how their colleagues tackle scientific problems and how to pool knowledge and talents to conduct interdisciplinary research. Perhaps equally important, the group learned to have fun at the same time.

"We decided to build a cell torture device instead," said Todd Thorsen of the Massachusetts Institute of Technology, who presented the group's findings to conference attendees.

Sequencing a Single Molecule of Protein

The study of protein structure and function is central to understanding living systems. However, the diversity and complexity of proteins render even the simplest characterizations challenging. The most basic level, determining the primary structure, involves sequencing the polypeptide chain. Even state-of-the-art commercial sequencing techniques require picomolar samples, equivalent to micrograms of protein or $\sim 10^{13}$ molecules. In contrast to this scale, laboratory experiments at the forefront of the field can access and manipulate single proteins with various physical techniques. These experiments have already shed light on structure and dynamics. Beyond simple sequencing, the higher-order structure of proteins—linked to understanding the folding process—remains elusive in the general case.

Problem

As typical methods for determining sequence and structure of proteins require large quantities of the molecule, these studies are often delayed until the requisite quantities are synthesized or purified. In the case of high-resolution crystallography, additional effort is required to crystallize sufficient quantities of the protein. Given the appearance of ground breaking single-protein studies with new tools, will it soon be possible to sequence a single molecule of protein? Consider a combination of existing techniques or newer techniques which need to be developed; for example:

1. Modifications of common amino acid sequencing techniques (filtration, cleavage, etc.)

2. Mass spectrometry
3. Optical tweezers
4. Cantilever-based force measurements
5. Nanopores/microfluidics
6. Scanning probe methods
7. Crystallography
8. Electron holography

Solution

The protein sequencing group considered themselves lucky. With a clearly defined problem in hand, several members came to the first session with ideas about what the solution should look like. Jotted down on hotel letterhead the night before, it was clear that several of these eleven men wanted their solution on the fast track to the final presentation. The thought of finishing early and taking off to Disneyland was considered, but was taken off the table when everyone began to realize that not everyone had the same answer to the problem.

The group's problem was to figure out how to sequence a single protein molecule. The sequencing techniques currently used by researchers require a large and highly concentrated sample of a protein. However, many proteins exist naturally in extremely small quantities: an individual cell may only have one or two copies of specific hormones and transcription factors. The ability to sequence these proteins would help in determining their structure; and, by combining sequence and structure information, large amounts of a specific protein could be produced and used in therapies. Such techniques could also be used diagnostically by identifying specific proteins associated with conditions or diseases.

There was little debate about the focus group's goal or its importance, but deciding what was the best line of attack proved challenging. There were many strong personalities; as a result, no one person was able to force his vision upon his colleagues. Some group members were in favour of "visualization" and believed that if a protein could be linearized and attached to a solid surface without any contamination then the sequence could be read using an atomic force microscope. Other members were in favour of the "flow channel" method in which a protein would be linearized and passed though a channel that would detect the sequence via nano-array. Someone else continued to stress the importance of using information available from the sequence of the human genome, by checking the determined amino acid sequence against known DNA sequences.

The differences in opinion regarding the ideal solution led to a tendency for individuals to interject with statements or questions that would pull the discussion toward the idea in which they were most interested. Even with this tug-of-war, most of the group's discussions were very productive and focused on specific aspects of the problem. Eventually an agreement was reached to focus on the solution that was showcased in the group's final presentation. It was not a coincidence that this solution involved input from most of the group.

The first decision to be made was whether the protein should remain intact, throughout the sequencing process, or if each amino acid should be systematically cleaved and detected. One of the benefits of keeping the sequence intact is that a single protein molecule could be sequenced many times. The group members in favour of visualization and the flow channel solutions cited repeatability as a huge benefit to their approaches. However, it was estimated that the visualization technique would take a trained technician an entire day to sequence a single protein, and a more efficient method was desired. The flow channel solution was also ruled out because most of the group was convinced that the forces exerted on the linearized protein as it passed though the channel would break it apart. So the group was forced to deal with the fact that the protein would need to be chopped up. There would be only one opportunity to read the sequence and then "game over."

The group decided that the chopping reactions currently used in sequencing could be used in single molecule sequencing as well. Using specific chemical reactions, individual amino acids could be cleaved from the amine end of the protein one at a time. In the first step of the group's design, the protein would be bound to the sample chamber at the carboxyl terminus. There was concern about losing the single protein molecule in this step, so it was suggested that a fluorescent probe could bind to the protein. Once detected, the stepwise reaction would cleave off a single amino acid to be identified.

The free amino acid would then be washed down into the first detection chamber. Here the amino acid would be temporarily bound to a silver substrate, and a laser would be used to generate a surface enhanced Raman spectra. SERS detection provides some information about the structure of a molecule, and may be able to determine the specific amino acid, but at a minimum could be used to ensure that an amino acid was released during the cleaving reaction, which only goes to 99.8 percent completion.

The amino acid would then be washed down into the second detection chamber called the riboswitch chamber. A riboswitch is a sequence of RNA that cuts itself in half when a specific molecule binds to it. Two naturally occurring riboswitches have been discovered that are tripped by glycine and lycine, respectively; and the group suspected that switches specific to the remaining 18 amino acids could be engineered. The riboswitches that are specific to each of the 20 amino acids would be attached along a wall of this detection chamber. Like balloons on strings, attached to each variety of riboswitch would be a specific coloured quantum-dot. Also called Qdots, these semiconductor nanocrystels light up in a variety of colours.

All of the lycine switches, for example, could be red; and all of the glycine switches could be green. When the amino acid enters the chamber it would bind to its specific riboswitch, which would then cleave itself; and a specific coloured Qdot would be released. A sensor would detect the colour, and the amino acid could be identified and then compared to the Raman spectra results. The amino acid and the Qdot are then washed out of the chamber and another amino acid could be cleaved in the sample chamber. The determined sequence would then be compared to known sequences in a database.

As creative and colourful as this idea is, the group identified several places where the mechanism could break down. One concern that deeply bothered the group was how to ensure that the amino acid did not get stuck to a wall on its way through the detection chamber. It was suggested that a solution with a high salt concentration could be used to wash off a stuck amino acid, but that could raise stability problems with the surface/ riboswitch/Qdot complexes.

Research challenges were identified at each of the steps in the group's solution. The sample chamber needs to be scaled down to a single molecule. The SERS detection chamber needs an optimized surface substrate, and Raman signature spectra for each of the 20 amino acids need to be determined. The riboswitch chamber needs switches and quantum dots specific to all 20 amino acids and a way to attach the switches to the chamber wall and the dots to the switches.

The second solution presented was the rejection of the central dogma. More dramatic and perhaps less practical of a solution, some members of the group hope to achieve "reverse translation." The argument presented was that though an enzyme that could achieve reverse translation has not been identified, it may exist somewhere in nature. After all, no one believed that reverse transcription was possible until it was discovered that nature had found a way. Even if the enzyme could not be found in nature right now, perhaps it could be created in a laboratory. In order to work, the enzyme would need to be able to use tRNA to identify each amino acid in order and ligate the RNA codons to generate the mRNA.

Another solution involved the riboswitch model. If DNA could be released instead of a quantum dot, then each piece of DNA representing an amino acid could be ligated to the previous piece, creating a sequence of DNA that corresponded to the amino acid sequence. The standard procedures for DNA sequencing could then be used and, in effect, reverse translation achieved.

What made this focus group unique was the specificity of the problem. Because there was no real debate about what the problem was, there was time to address several possible solutions and their individual challenges. The design and the discussions were about details.

12

Atomic Orbitals and Carbon Nanotubes

Atomic orbitals are the basis for the formation of *chemical bonds*, i.e., for the formation of molecules and solids. Thanks to the modern nanotechnological tools the atomic orbitals are no longer merely a mathematical abstraction, but something very real that can be observed directly. We begin with a short introduction to hydrogen-like atoms, and then move on to carbon atoms, graphene sheets, and finally the carbon nanotube molecules.

Schrodinger Equation for Hydrogen-like Atoms

In the following we analyze one-electron hydrogen-like atoms like hydrogen itself, H, the helium-I ion, He^+, and the lithium-II ion, Li^{++}, and our starting point is therefore the *Schrodinger equation* for an electron moving in the *Coulomb potential* from a heavy nucleus at rest with the positive charge $+Ze$.

We consider an infinitely heavy nucleus with charge $+Ze$ placed in the center of our coordinate system. The electrostatic potential energy $V(r)$ for an electron (with charge $-e$ and mass m_e) moving around such a nucleus is given by

$$V(r) = \frac{\alpha}{r}, \text{ with } \alpha = \frac{Ze^2}{4\pi\varepsilon_0}. \qquad ...(1)$$

This potential energy depends only on the distance r between the electron and the nucleus and not on any angles. It is therefore natural to work with spherical coordinates (r, θ, ϕ) instead of the usual cartesian coordinates (x, y, z). These two sets of coordinates are related by

$$x = r \sin\theta \cos\phi,$$
$$y = r \sin\theta \sin\phi, \quad ...(2)$$
$$z = r \cos\theta.$$

Substituting as usual the classical momentum p with the differential operator $\frac{\hbar}{i}\nabla$, the classical energy equation for the electron $E = p^2/2m_e - \alpha/r$ becomes the time-independent Schrodinger equation

$$\left[-\frac{\hbar^2}{2m_e}\nabla^2 - \frac{\alpha}{r}\right]\psi_E(r) = E\psi_E(r). \quad ...(3)$$

In spherical coordinates ∇^2, and the Schrodinger equation takes the following form

$$\left[-\frac{\hbar^2}{2m_e}\left\{\frac{1}{r^2}\frac{\partial}{\partial r}\left(r^2\frac{\partial}{\partial r}\right) + \frac{1}{r^2\sin\theta}\frac{\partial}{\partial\theta}\left(\sin\theta\frac{\partial}{\partial\theta}\right) + \frac{1}{r^2\sin^2\theta}\frac{\partial^2}{\partial\phi^2}\right\} - \frac{\alpha}{r}\right]$$
$$\psi_E(r) = E\psi_E(r). \quad ...(4)$$

It seems to be a very difficult affair to find the eigenstates $\psi_E(r)$ that solve this equation. But the spherical coordinates make things easier. Fortunately, the complicated differential operator acting on $\psi_E(r)$ on the left-hand side of the equation has very distinct parts acting on each of the coordinates r, θ, and ϕ, respectively. We therefore separate the wavefunction as a product of functions of each of the three coordinates,

$$\psi_E(r) = R(r)\,\Theta(\theta)\,\Phi(\phi), \quad ...(5)$$

and find simple solutions for each of these functions.

Azimuthal Functions $\Phi_m(\phi)$

We begin with the *azimuthal angle* ϕ. The only place where the differential operator in Eq. (4) acts on ϕ is the term containing $\frac{\partial^2}{\partial\phi^2}$. Hence we seek a function $\Phi_m(\phi)$ that gives a simple result when acted upon by this differential operator:

$$\frac{\partial^2\Phi_m(\phi)}{\partial\phi^2} = -m^2\Phi_m(\phi), \quad ...(6)$$

where m is a constant (maybe complex) to be determined. We see immediately that $\Phi_m(\phi) = \exp(im\phi)$ solves Eq. (6), but what about the boundary condition? Naturally, we would like the wavefunction to be single-valued, so upon changing the azimuthal angle by 2π when should obtain the same value, i.e. $\Phi_m(\phi + 2\pi) = \Phi_m(\phi)$. But with $\Phi_m(\phi) = \exp(im\phi)$ this is only possible for integer values of m. We therefore obtain the solution

$$\Phi_m(\phi) = \frac{1}{\sqrt{2\pi}} e^{im\phi}, \quad m = 0, \pm 1, \pm 2, \ldots \qquad \ldots(7)$$

The prefactor ensures proper normalization:

$$\int_0^{2\pi} d\phi \left|\Phi_m(\phi)\right|^2 = 1.$$

Polar Functions $\boldsymbol{\Theta_{lm}(\theta)}$

We now choose one particular integer value of m and keep it fixed. When this is done $\frac{\partial^2}{\partial \phi^2}$ simply gets replaced by the number $-m^2$. We then focus on the θ-dependent part of the *Schrodinger equation* and seek functions $\Theta_{lm}(\theta)$ that gives a simple result when the θ-part of the differential operator acts on it:

$$\left[\frac{1}{\sin\theta}\frac{\partial}{\partial\theta}\left(\sin\theta\frac{\partial}{\partial\theta}\right) - \frac{m^2}{\sin^2\theta}\right]\Theta_{lm}(\theta) = -l(l+1)\Theta_{lm}(\theta). \qquad \ldots(8)$$

Apriori, the constant $l(l + 1)$ can be any complex number, but anticipating the solution we have written it as seen. The solution of Eq. (8) is found by changing variables from θ to $\mu \equiv \cos\theta$, whereby the equation for $\Theta_{lm}(\theta) = P_l^m(\mu)$, becomes

$$\frac{d}{d\mu}\left[(1-\mu^2)\frac{dP_l^m(\mu)}{d\mu}\right] - \frac{m^2}{1-\mu^2}P_l^m(\mu) = -l(l+1)P_l^m(\mu), \quad \mu = \cos\theta. \qquad \ldots(9)$$

This is one of the classical differential equation of mathematical analysis. Given the boundary condition $-1 \le \mu \le 1$ the solutions are known to be the so-called associated Legendre polynomials $P_l^m(\mu)$, where l is a non-negative integer, and where $l \ge |m| \ge 0$:

$$P_l^m(\mu) = \frac{(-1)^m}{2^l\, l!}(1-\mu^2)^{\frac{m}{2}}\frac{d^{l+m}}{d\mu^{l+m}}(\mu^2-1)^l, \quad \begin{cases} l = 0,1,2,\ldots \\ m = -l,\ldots,l. \end{cases} \qquad \ldots(10)$$

With proper normalization, $\int_{-1}^{1} d(\cos\theta)\left|\Theta_{lm}(\theta)\right|^2 = 1$ we arrive at

$$\Theta_{lm}(\theta) = \sqrt{\frac{(l+\frac{1}{2})(i-m)!}{(l+m)!}}P_l^m(\cos\theta), \quad \begin{cases} l = 0,1,2,\ldots \\ m = -l,-(l-1),\ldots(l-1),l. \end{cases} \qquad \ldots(11)$$

We give some explicit expressions for this wavefunction in the next subsection.

Spherical Harmonics $Y_l^m(\theta,\phi) = \Theta_{lm}(\theta)\,\Phi_m(\phi)$

We have now solved the angular part of the problem. For reasons it is costumary to combine the two angular functions $\Theta_{lm}(\theta)$ and $\Phi_m(\phi)$ into one, the so-called spherical harmonic Y_l^m (θ, ϕ):

$$Y_l^m(\theta,\phi) = \Theta_{lm}(\theta)\Phi_m(\phi), \quad \begin{cases} l = 0,1,2,\ldots \\ m = -l,-(l-1),\ldots,(l-1),l. \end{cases} \qquad \ldots(12)$$

The *spherical harmonics* are eigenfunctions to the angular part of the Laplace operator ∇^2,

$$\left[\frac{1}{\sin\theta}\frac{\partial}{\partial\theta}\left(\sin\theta\frac{\partial}{\partial\theta}\right) + \frac{1}{\sin^2\theta}\frac{\partial^2}{\partial\phi^2}\right] Y_l^m(\theta,\phi) = -l(l+1)\,Y_l^m(\theta,\phi),$$

$$\begin{cases} l = 0,1,2,\ldots \\ m = -l,\ldots,l. \end{cases} \qquad \ldots(13)$$

Note that eventhough m is an index in Y_l^m (θ, ϕ) it does not appear in the eigenvalue $-l(l + 1)$. Below we list all spherical harmonics with $l \leq 3$.

$$Y_0^0 = \sqrt{\frac{1}{4\pi}}$$

$$Y_3^3 = -\sqrt{\frac{35}{64\pi}}\sin^3\theta\, e^{i3\phi}$$

$$Y_3^2 = \sqrt{\frac{105}{32\pi}}\sin^2\theta\cos\theta\, e^{i2\phi}$$

$$Y_3^1 = -\sqrt{\frac{21}{64\pi}}\sin\theta(5\cos^2-1)\, e^{i\phi}$$

$$Y_3^0 = \sqrt{\frac{7}{16\pi}}(5\cos^3\theta - 3\cos\theta)$$

$$Y_3^{-1} = \sqrt{\frac{21}{64\pi}}\sin\theta(5\cos^2-1)\, e^{-i\phi}$$

$$Y_3^{-2} = \sqrt{\frac{105}{32\pi}}\sin^2\theta\cos\theta\, e^{-i2\phi}$$

$$Y_3^{-3} = \sqrt{\frac{35}{64\pi}}\sin^3\theta\, e^{-i3\phi}$$

$$Y_1^1 = -\sqrt{\frac{3}{8\pi}}\sin\theta\, e^{i\phi}$$

$$Y_1^0 = \sqrt{\frac{3}{4\pi}}\cos\theta$$

$$Y_1^{-1} = \sqrt{\frac{3}{8\pi}}\sin\theta\, e^{-i\phi}$$

$$Y_2^2 = \sqrt{\frac{15}{32\pi}}\sin^2\theta\, e^{i2\phi}$$

$$Y_2^1 = -\sqrt{\frac{15}{8\pi}}\cos\theta\cos\theta\, e^{i\phi}$$

$$Y_2^0 = \sqrt{\frac{5}{16\pi}}(3\cos^2\theta - 1)$$

$$Y_2^{-1} = \sqrt{\frac{15}{8\pi}}\sin\theta\cos\theta\, e^{-i\phi}$$

$$Y_2^{-2} = \sqrt{\frac{15}{32\pi}}\sin^2\theta\, e^{-i2\phi}$$

$$\ldots(14)$$

Radial Functions $R_{nl}(r)$

The last part of the *Schrodinger equation* Eq. (4) to be solved concerns the radial coordinate r. We plug in $\psi_E(r) = R(r)\, Y_l^m(\theta, \phi)$, use the eigenvalue equation Eq. (13), and divide out the spherical harmonic $Y_l^m(\theta, \phi)$. The result is the following equation for $R(r)$,

$$\frac{d^2R(r)}{dr^2} + \frac{2}{r}\frac{dr(r)}{dr} - \frac{l(l+1)}{r^2}R(r) + \frac{2m_e}{\hbar^2}\left[E + \frac{\alpha}{r}\right]R(r) = 0. \quad ...(15)$$

This differential equation is then made dimensionless. Since the equation contains the parameters m_e, $\hbar$, and $\alpha = Ze^2/4\pi\varepsilon_0$ we form the characteristic length a_Z and the characteristic energy E_Z (the index Z refers to the charge $+Ze$ of the nucleus),

$$a_Z = \frac{\hbar^2}{m_e\alpha} \quad ...(16a)$$

$$E_Z = \frac{m_e\alpha^2}{\hbar^2} \quad ...(16b)$$

In analogy with the parameter $l(l + 1)$ for Θ_{lm} we now introduce the dimensionless (and perhaps complex) parameter n, and use that together with a_Z and E_Z to define dimensionless measures of energy, n, and of length, ρ:

$$n = \sqrt{\frac{E_Z}{2|E|}}, \Rightarrow E = \frac{-1}{2n^2}E_Z, \quad ...(17a)$$

$$\rho = \frac{2}{n}\frac{r}{a_Z}, \Rightarrow r = \frac{n}{2}\rho\, a_Z. \quad ...(17b)$$

For negative energies in terms of ρ and n the radial equation Eq. (15) becomes

$$\frac{d^2R_{nl}(\rho)}{d\rho^2} + \frac{2}{\rho}\frac{dR_{nl}(\rho)}{d\rho} + \left[-\frac{1}{4} + \frac{n}{\rho^2} - \frac{l(l+1)}{\rho^2}\right]R_{nl}(\rho) = 0. \quad ...(18)$$

This equation is solved by studying the asymptotic behaviour for $\rho \to \infty$ and $\rho \to 0$. In the limit $\rho \to \infty$ the terms in Eq. (18) containing $1/\rho$ and $1/\rho^2$ drop out, and thus

$$\frac{d^2R_{nl}(\rho)}{d\rho^2} - \frac{1}{4}R_{nl}(\rho) \approx 0 \Rightarrow R_{nl}(\rho) \approx e^{-\frac{1}{2}\rho}. \quad ...(19)$$

We have chosen the exponentially decaying solution since the probability of the electron to be infinitely far away from the nucleus must vanish. In the other limit, $\rho \to 0$ we keep the derivatives of R as well as the dominant term $l(l + 1)/\rho^2$:

$$\frac{d^2 R_{nl}(\rho)}{d\rho^2} + \frac{2}{\rho}\frac{dR_{nl}(\rho)}{d\rho} - \frac{l(l+1)}{\rho^2} R_{nl}(\rho) \approx 0 \Rightarrow R_{nl}(\rho) \approx \rho^l, \quad ...(20)$$

where we have chosen the solution that remains finite at $\rho = 0$. We thus end with the following form of $R_{nl}(\rho)$:

$$R_{nl}(\rho) = w_{nl}(\rho)\rho^l e^{-\frac{1}{2}\rho}. \quad ...(21)$$

Upon insertion of Eq. (21) into Eq. (18) we arrive at

$$\rho\, w''_{nl} + (2l + 2 - \rho)w'_{nl} + (n - l - 1)w_{nl} = 0. \quad ...(22)$$

To ensure the exponential decay for $\rho \to \infty$ given in Eq. (21) $w(\rho)$ must be a polynomium of finite degree, and from this follows that n must be an integer subject to the condition $n \geq l + 1$. Such polynomial solutions to the differential equation Eq. (22) are known to be the so-called generalized *Laguerre polynomials* L_{n+l}^{2l+1},

$$L_n^p(\rho) = \frac{(-1)^p (n!)^2}{p!(n-p)!}\rho^{-p} e^{\rho} \frac{d^{n-p}}{d\rho^{n-p}}(\rho^n e^{-\rho}), \; 0 \leq p \leq n. \quad ...(23)$$

To ensure the normalization condition $\int_0^\infty |R_{nl}(\rho)|^2 \rho^2 dp = 1$ we find the radial wavefunction $R_{nl}(\rho)$ to be

$$R_{nl}(\rho) = -\frac{n}{4}\sqrt{\frac{(n-l-1)!}{[(n+l)!]^3}}\rho^l L_{n+l}^{2l+1}(\rho)\, e^{-\frac{1}{2}\rho}, \quad \begin{cases} n = 1,2,3,\ldots \\ n \geq l+1. \end{cases} \quad ...(24)$$

Below we list the radial wavefunctions $R_{nl}(r)$ for $n = 1, 2, 3$ and $l = 0, 1, \ldots, n - 1$. We have reinstated the actual radial coordinate r.

$$R_{10}(r) = \frac{2}{\sqrt{a_Z^3}}\exp\left(-\frac{r}{a_Z}\right)$$

$$R_{20}(r) = \frac{1}{\sqrt{2a_Z^3}}\left(1 - \frac{1}{2}\frac{r}{a_Z}\right)\exp\left(-\frac{r}{2a_Z}\right)$$

$$R_{21}(r) = \frac{1}{2\sqrt{6a_Z^3}}\left(\frac{r}{a_Z}\right)\exp\left(-\frac{r}{2a_Z}\right) \quad ...(25)$$

$$R_{30}(r) = \frac{2}{3\sqrt{3a_Z^3}}\left(1 - \frac{2}{3}\frac{r}{a_Z} - \frac{2}{27}\frac{r^2}{a_Z^2}\right)\exp\left(-\frac{r}{3a_Z}\right)$$

$$R_{31}(r) = \frac{8}{27\sqrt{6a_Z^3}}\left(\frac{r}{a_Z} - \frac{1}{6}\frac{r^2}{a_Z^2}\right)\exp\left(-\frac{r}{3a_Z}\right)$$

$$R_{32}(r) = \frac{4}{81\sqrt{30a_Z^3}}\left(\frac{r^2}{a_Z^2}\right)\exp\left(-\frac{r}{3a_Z}\right)$$

Energies and Sizes of the Atomic Orbitals

Returning to the normal units, we find the following energies E_n of the atomic orbitals,

$$E_n = -\frac{1}{2n^2}E_Z = -\frac{m_e\alpha^2}{2\hbar^2}\frac{1}{n^2} \Rightarrow E_n = -\frac{m_e Z^2 e^4}{32\pi^2\varepsilon_0^2\hbar^2}\frac{1}{n^2},\ n = 1,2,3,\ldots \qquad \ldots(26)$$

Note that n is the only quantum number entering the expression for the eigenenergies. Each energy level therefore has a degeneracy given by the number allowed values of l and m for a given n. Since $l = 0, 1, \ldots (n-1)$ and $m = -l, -(l-1), \ldots, (l-1), l$ we have

$$\text{degeneracy of } E_n = \sum_{l=0}^{n-1}(2l+1) = n^2. \qquad \ldots(27)$$

The size of the atomic orbitals is not well defined due to the statistical nature of the wavefunction. It is only characterized fully by calculating all the expectation values $\langle r^k\rangle$ for any power k. Since the potential behaves as $1/r$ it is natural to estimate the size of the orbitals by calculating $\langle 1/r\rangle$,

$$\left\langle\frac{1}{r}\right\rangle_{nlm} = \int_0^\infty dr\, r^2\frac{1}{r}|R_{nl}(r)|^2 = \frac{1}{n^2}\frac{1}{a_Z}. \qquad \ldots(28)$$

For the ground state ψ_{100} of the hydrogen atom we find (with $Z = 1$) the ground state energy E_0 and the size a_0 to be

$$a_0 = a_{Z=1} = \frac{4\pi\varepsilon_0\hbar^2}{m_e e^2} = 0.053\text{ nm}, \qquad \ldots(29a)$$

$$E_0 = E_{Z=1} = -\frac{m_e e^4}{32\pi^2\varepsilon_0^2\hbar^2} = -13.6\text{ eV}. \qquad \ldots(29b)$$

Atomic Orbitals: Shape and Nomenclature

The shape of some of the atomic orbitals $\psi_{nlm}(r)$ in hydrogen-like atoms has been shown in the figure. The figure contains all orbitals with $n \leq 4$ with the omission of negative values of m since the orbitals (n, l, m) and $(n, l, -m)$ have the same shape.

When specifying an orbital it is standard to write the value for n followed by a code letter for the value of l according to the following table

value of l :	0	1	2	3	4	
code letter :	s	p	d	f	g	...(30)

The value of m can be given as a subscript to the code letter for l, but often it is omitted, and instead one writes the number of electrons with the same value of n and l as a superscript to the code letter for l,

$$\text{orbital term} = (n \text{ value})(l \text{ code letter})^{(\text{number of electrons})}. \quad ...(31)$$

As an example of the nomenclature we take the carbon atom. In its ground state it has two electrons (one spin up and one spin down) with $(n, l) = (1, 0)$, two with $(n, l) = (2, 0)$, and two with $(n, l) = (2, 1)$. Thus the electron configuration for carbon is written as

$$[\text{C}] = 1s^2 2s^2 2p^2 = [\text{He}]2s^2 2p^2, \quad ...(32)$$

where we also have use the electron configuration of He, the noble gas in the periodic table before C. This latter notation is helpful, since it emphasizes the orbitals of the valence electrons and just summarizes the inner core electrons as the filled electron orbitals of a nobel gas.

Angular Momentum: Interpretation of *l* and *m*

The angular part of the Laplacian ∇^2 in spherical coordinates is closely related to the angular momentum operator $\hat{L}$. In quantum mechanics $\hat{L}$ is defined by

$$\hat{L} = r \times \hat{p} = \frac{\hbar}{i}\begin{pmatrix} y\frac{\partial}{\partial z} - z\frac{\partial}{\partial y} \\ z\frac{\partial}{\partial x} - x\frac{\partial}{\partial z} \\ x\frac{\partial}{\partial y} - y\frac{\partial}{\partial x} \end{pmatrix}. \quad ...(33)$$

Transforming to spherical coordinates yields

$$\hat{L} = \begin{pmatrix} \hat{L}_x \\ \hat{L}_y \\ \hat{L}_z \end{pmatrix} = \frac{\hbar}{i}\begin{pmatrix} -\sin\phi - \cot\theta\cos\phi\frac{\partial}{\partial\phi} \\ +\cos\phi\frac{\partial}{\partial\theta} - \cot\theta\sin\phi\frac{\partial}{\partial\phi} \\ \frac{\partial}{\partial\phi} \end{pmatrix}. \quad ...(34)$$

From this follows by straightforward algebra that

$$\hat{L}^2 = \hat{L}_x^2 + \hat{L}_y^2 + \hat{L}_z^2 = \frac{1}{\sin\theta}\frac{\partial}{\partial\theta}\left(\sin\theta\frac{\partial}{\partial\theta}\right) + \frac{1}{\sin^2\theta}\frac{\partial^2}{\partial\phi^2}. \quad ...(35)$$

But this is noting but the angular part of the Laplacian ∇^2, and combining Eqs. (12) and (13) with Eqs. (34) and (35) leads to the following important results

$$\hat{L}_z Y_l^m(\theta,\phi) = \hbar m Y_l^m(\theta,\phi) \quad ...(36)$$

$$\hat{L}^2 Y_l^m(\theta,\phi) = \hbar^2 l(l+1) Y_l^m(\theta,\phi). \quad ...(37)$$

Consequently, for the wavefunction $Y_l^m(\theta, \phi)$ we can identify $\sqrt{l(l+1)}\hbar$ with the absolute size of the angular momentum, and $m\hbar$ with its z-component L_z.

Since one cannot determine the position r and the momentum $\hat{p}$ simultaneously with absolute certainty, it is expected that one cannot determine all three components of the angular momentum $\hat{L} = r \times \hat{p}$ simultaneously with absolute certainty. This is in fact true. As seen in Eqs. (36) and (37) it is possible to determine the values of $\hat{L}_z$ and $\hat{L}$ simultaneously. But we also note that even if $\hat{L}_z$ takes its maximal value $l\hbar$ it is still smaller than the total size $\sqrt{l(l+1)}$ of $|\hat{L}|$,

$$\max\{\hat{L}_z\} = l\hbar < \sqrt{l(l+1)}\hbar = |\hat{L}|. \quad ...(38)$$

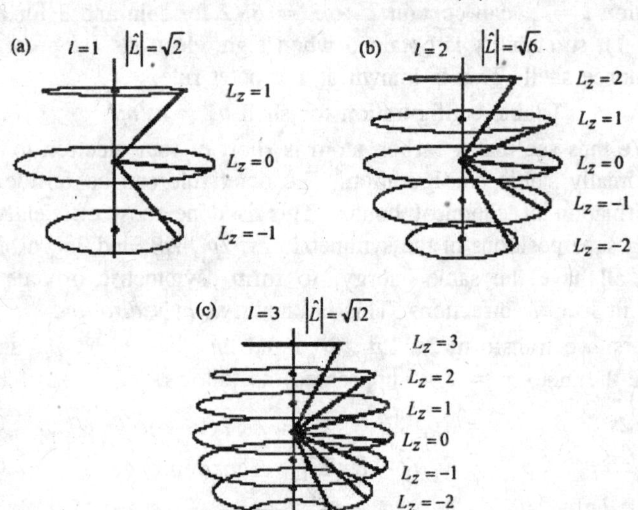

Fig. 12.1. Illustration of the precessing angular momentum for l=1,2,3 and m=–l,–(l–1),..., (l–1), l.

We can conclude that there is always some angular momentum left in the *x* and *y* direction. This can be illustrated by imagining that the $\hat{L}$-vector precesses around the *z*-axis with a fixed *z*-component $\hat{L}_z$.

Carbon Atom and sp^2 Hybridization

The ability of a single carbon atom to form up to four strong covalent bonds makes it a key player in organic chemistry and hence a corner stone in the formation of biological tissue. Moreover, the discoveries of the C_{60} "*bucky ball*" and the carbon nanotube molecules in 1986 and 1992, respectively, have given carbon a status in nanoscience and nanotechnology comparable to that of silicon in *microtechnology*. We therefore give a brief overview of *carbon* and *carbon nanotubes* in the following sections.

Carbon is the sixth element in the periodic table. The nucleus of the most abundant carbon isotope contains *six protons* and *six neutrons*, while the electron configuration is given by $1s^2 2s^2 2p^2$ as stated in Eq. (32). Of these six electrons the four in the 2*s* and 2*p* orbitals are valence electrons able to participate in the formation of chemical bonds.

It is a well known fact that atoms are most stable when the *s* and *p* orbitals of the outer shell are occupied. An *s* orbital can contain two electrons (spin up and spin down), while a *p* orbital, where per definition $l = 1$, can contain $2 \times 3 = 6$ (2 for spin and 3 for $m = -1, 0, 1$), so stability is obtained when eight electrons are present in the valence shell. This is known as the octet rule,

$$[\text{stable configuration for shell } n] = ns^2np^6. \qquad \ldots(39)$$

We thus see that a carbon atom is short of four electron to have a maximally stable configuration. The octet rule can be fulfilled by the formation of chemical bonds. This is done most efficiently by making superpositions of the symmetric 2*s*, $2p_1$, $2p_0$, and $2p_{-1}$ orbitals, which all have the same energy, to form asymmetric orbitals that points in specific directions. This is known as *hybridization*.

First we transform 2*s*, $2p_1$, $2p_0$, and $2p_{-1}$ into *s*, p_x, p_y, and p_z (where the index $n = 2$ is suppressed) as follows,

$$s(r) \equiv 2s \qquad = R_{20}Y_0^0 \qquad = A\sqrt{2}(1-\rho)e^{-\rho}, \qquad \ldots(40a)$$

$$p_x(r) \equiv 2p_1 - 2p_{-1} = R_{21}Y_1^1 - R_{21}Y_{-1}^1 = A\cos\phi\sin\theta\, pe^{-\rho}, \qquad \ldots(40b)$$

$$p_y(r) \equiv 2p_1 + 2p_{-1} = R_{21}Y_1^1 + R_{21}Y_{-1}^1 = A\sin\phi\sin\theta\, pe^{-\rho}, \qquad \ldots(40c)$$

$$p_z(r) \equiv 2p_0 \qquad = R_{21}Y_1^0 \qquad = \sqrt{6}A\cos\theta(1-\rho)e^{-\rho}, \qquad \ldots(40d)$$

where A is a constant and ρ is the dimensionless length from Eq. (17b).

We can now define the so-called sp^2 *hybridization*. It consists of the following superpositions of $s(r)$, $p_x(r)$, and $p_y(r)$, i.e., superpositions involving one s and two p functions (leaving p_z as it is),

$$sp_a^2(r) \equiv \sqrt{\frac{4}{6}}s(r) + \sqrt{\frac{2}{6}}p_x(r), \qquad \text{...(41a)}$$

$$sp_b^2 \equiv \sqrt{\frac{2}{6}}s(r) - \sqrt{\frac{1}{6}}p_x(r) + \sqrt{\frac{3}{6}}p_y(r), \qquad \text{...(41b)}$$

$$sp_c^2(r) \equiv \sqrt{\frac{2}{6}}s(r) - \sqrt{\frac{1}{6}}p_x(r) - \sqrt{\frac{3}{6}}p_y(r). \qquad \text{...(41c)}$$

The shape of the three hybridized sp^2 orbitals is shown in figure. Here in panels (d), (e), and (f) it is seen how $sp_a^{\ 2}$ points in the x direction, while $sp_b^{\ 2}$ and $sp_c^{\ 2}$ are rotated $+120°$ and $-120°$ in the xy plane, respectively. The reason for this result lies in the fact that p_x is positive for $x > 0$ and negative for $x < 0$. When adding this angular dependent function to the rotational invariant function s, we obtain the assymetric results.

The p_z orbital does not enter the sp^2 hybridization, and it remains a figure eight-like orbital perpendicular to the xy plane containing the three sp^2 orbitals built from s, p_x, and p_y. In figure, two 3D sketches of the resulting sp^2 hybridized orbitals and the non-hybridized orbital p_z.

We only study sp^2 hybridization here, since this is relevant to the formation of graphene planes and carbon nanotubes. However, part of the rich organic chemistry of carbon relates to the fact that also sp^1 and sp^3 can occur. In sp^1 hybridization s and p_x hybridizes into $sp_x^{\ 1}$ leaving p_y and p_z perpendicular to sp^1 and each other. The uni-directional sp^1 bond is found in compounds containing the linear carbon tripple bond –C≡C–. In sp^3 hybridization all four valence orbitals hybridize an form the characteristic tetragonal structure that each carbon atoms has in the diamond structure.

Graphene, Sigma and Pi Bonds

As mentioned the formation of chemical bonds is one way for an atom to fulfil the octet rule. It is a general fact of quantum mechanics, that when an orbital from one atom is in close proximity of an orbital from another atom, two new hybridized orbitals can be formed by superposition. We shall use this description to explain the graphene structure, which is a planar hexagonal lattice of carbon atoms. Graphene is the basis for understanding carbon nanotubes.

The following discussion is a simplification of the real quantum orbital calculations. To be specific, consider two carbon atoms 1 and 2. Let atom 1 have the sp^2 orbitals have the orientations $sp^2_{a1} \sim 0°$, $sp^2_{b1} \sim +120°$, and $sp^2_{c1} \sim -120°$, and let atom 2 be in the mirror inverted state such that $sp^2_{a2} \sim 180°$, $sp^2_{b2} \sim -60°$, and $sp^2_{c2} \sim +60°$.

The two orbitals sp^2_{a1} and sp^2_{a2} points towards each other. When the two carb on atoms therefore are close enough, they hybridize and form the two sigma-bonds σ and σ^* with low and high energy, respectively, defined as follows

$$\sigma(r) = \frac{1}{\sqrt{2}}\left[sp^2_{a1}(r) + sp^2_{a2}(r)\right] \text{ very low energy,} \qquad ...(42)$$

$$\sigma^*(r) = \frac{1}{\sqrt{2}}\left[sp^2_{a1}(r) - sp^2_{a2}(r)\right] \text{ very high energy.} \qquad ...(43)$$

The σ bond is the *strong covalent bond* that makes the carbon compounds very stable. The σ^* orbital has such a high energy that the electrons cannot be excited up into this orbital, it has thus a negligible influence on the molecule and can be disregarded. Hence, it is always the σ bond that matters. All sp^2 orbitals can pairwise form a σ-bond (and an inaccessible σ^*-orbital) with another atom. Each σ bond is always fully occupied with two electrons (*spin up and down*), one from each atom. The three σ bonds thus accounts for three of the four valence electrons of the carbon atom.

Similarly, the p_{z1} and p_{z2} orbitals from the two carbon atoms can form a superposition. However, since they are parallel to each other their spatial overlap is not as large as the two sp_a^2 orbitals that pointed towards each other, so the energies involved are much less. The hybridized bonds are called *π-bonds*,

$$\pi(r) = \frac{1}{\sqrt{2}}\left[p_{z1}(r) + p_{z2}(r)\right] \text{ slightly lower energy,} \qquad ...(44)$$

$$\pi^*(r) = \frac{1}{\sqrt{2}}\left[p_{z1}(r) - p_{z2}(r)\right] \text{ slightly higher energy.} \qquad ...(45)$$

The small energy difference means that while π orbital is the ground state, the π^* orbital is still accessible to the electrons, thus one expects (and finds) fast electron transfer (equal to large electrical conductivity) between the π orbitals in carbon compounds. The π orbital can hold two electrons (spin up and down), one from each carbon atom. This accounts for the last valence electron.

To summarize: the strength of the carbon compounds is due to the very strong and fully occupied σ bonds that lies in the xy plane. These bonds are electrically inert because the unoccupied σ^* orbitals have too high an energy to be reached. The rather high electrical conductivity of certain carbon compounds is due to the π orbitals because the unoccupied π^* bonds have only a slightly higher energy, and can therefore be reached.

It is now very easy to describe graphene, the planar hexagonal lattice of carbon atoms, that upon stacking forms graphite. In Fig. 12.2(a) is shown a the atomic structure of such a planar sheet of graphene. The small circular disks represent carbon atoms. The lines between the disks represent the σ bonds. The characteristic 120° angles are clearly recognized in the regular lattice. The mechanical in-plane strength of graphene is large due to the strong σ bonds.

While not shown in Fig. 12.2(a) there is also the π orbitals forming a cloud of electrons just above and just below the plane. Following the previous discussion about the occupied π orbitals and the unoccupied

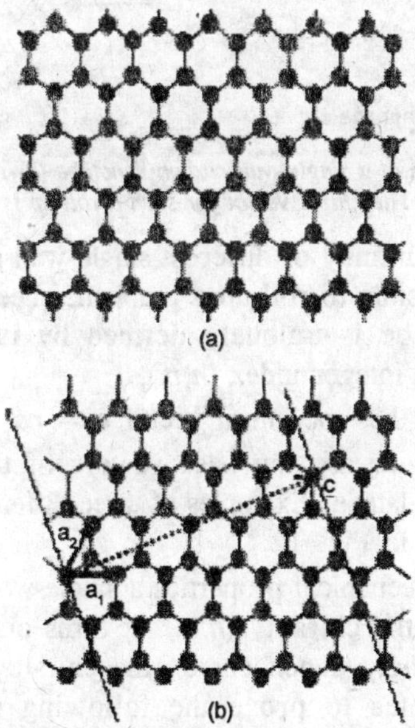

Fig. 12.2. (a) The planar hexagonal lattice of carbon atoms forming a graphene sheet. (b) The base vectors a_1 and a_2 defining the unit cell of the hexagonal lattice.

π^* orbitals, it is clear that graphene has a very good electrical conductivity parallel to the plane.

Graphite is formed by stacking a huge number of graphene sheets on top of each other, and consequently it is a highly anisotropic material. Graphite yields easily to shear stresses, since each graphene sheet is only weakly bound to each other. The lack of free electron orbitals make van der Waals forces the only binding perpendicular to the sheets. Electrically graphite is a poor conductor perpendicular to the graphene sheets, but it has a decent conductivity along the sheets.

Carbon Nanotubes

Carbon nanotube are extraordinary macromolecules containing only carbon. They are formed by rolling up *graphene sheets*. There exist two categories of nanotubes, the *single-wall nanotube* (*SWNT*) and *multi-wall nanotubes* (*MWNT*), which consist of several concentric single-wall nanotubes. In experiments one often finds that the nanotubes collect in so-called *bundles* or *ropes*.

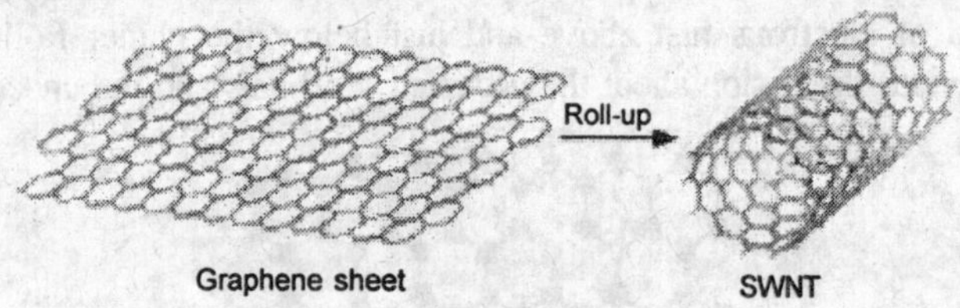

Fig. 12.3. The formation of a single-wall carbon nanotube (SWNT) by rolling up a sheet of graphene. The chiral vector for such a roll-up is defined in Fig. 12.2(b).

There exist an infinity of different single-wall nanotube the reason being that it is possible to roll-up a graphene sheet in many ways. A single-wall nanotube is uniquely defined by its chiral vector or equivalently its two-integer index (n, m),

nanotube (n,m) has the chiral vector $c = na_1 + ma_2$, ...(46)

where *textbf*a_1 and a_2 are the base vectors of the unit cell in the hexagonal graphene lattice. Examples of three different carbon nanotube structures are seen in Fig. 12.4.

Whereas the mechanical properties a single-wall carbon nanotubes is only weakly on the chirality (n,m), it turns out that the electrical properties are strongly dependent on the chirality. It is beyond the scope of these notes to prove the following result, but it is a straightforward, rather tedious, exercise in single-electron tight binding theory,

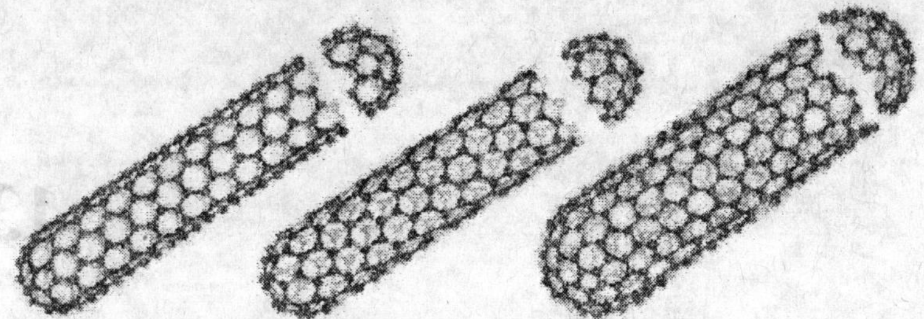

Fig. 12.4. The three single-wall carbon nanotubes (5,5), (9,0), and (10,5).

$$n - m = 3 \times \text{integer} \Rightarrow \text{metalic}$$

$$n - m \neq 3 \times \text{integer} \Rightarrow \text{semiconduc ting.} \quad \text{...(47)}$$

The diameter d and the chiral angle θ of a single-wall carbon nanotube are given by the C=C bond length a and the chiral index (n, m) by

$$d = \frac{a}{\pi}\sqrt{n^2 + m^2 + nm}, \quad \text{...(48)}$$

$$\tan\theta = \frac{\sqrt{3}m}{2n + m}. \quad \text{...(49)}$$

Table 12.1. Some physical parameters of single-wall carbon nanotubes.

Typical density	1400 kg/m³	Typical diameter	1.3 nm
Max tensile strength	30 GPa	Bond length C=C	0.142 nm
Young's modulus	1000 GPa	Optical gap	0.5 eV
Thermal conductivity	2000 W/m/K	Max current density	10^{13} A/m²

13

NANOTUBES

Nearly 10 years ago Sumio Iijima, first noticed odd *nanoscopic threads* lying in a smear of soot. Made of pure carbon, as regular and symmetric as crystals, these exquisitely thin, impressively long macromolecules soon became known as *nanotubes*, and they have been the object of intense scientific study ever since.

Just recently, they have become a subject for engineering as well. Many of the extraordinary properties attributed to *nanotubes*—among them, *superlative resilience*, *tensile strength* and *thermal stability*—have fed fantastic predictions of microscopic robots, dent-resistant car bodies and earthquake-resistant buildings. The first products to use nanotubes, however, exploit none of these. Instead the earliest applications are electrical. Some General Motors cars already include plastic parts to which nanotubes were added; such plastic can be electrified during painting so that the paint will stick more readily. And two nanotube-based lighting and display products are well on their way to market.

In the long term, perhaps the most valuable applications will take further advantage of *nanotubes*' unique electronic properties. Carbon nanotubes can in principle play the same role as silicon does in electronic circuits, but at a molecular scale where silicon and other standard semiconductors cease to work. Although the electronics industry is already pushing the critical dimensions of transistors in commercial chips below 200 nanometers (billionths of a meter)—about 400 atoms wide—engineers face large obstacles in continuing this miniaturization. Within this decade, the materials and processes on which the computer revolution has been built will begin to hit fundamental physical limits.

Still, there are huge economic incentives to shrink devices further, because the speed, density and efficiency of microelectronic devices all rise rapidly as the minimum feature size decreases. Experiments over the past several years have given researchers hope that wires and functional devices tens of nanometers or smaller in size could be made from nanotubes and incorporated into electronic circuits that work far faster and on much less power than those existing today.

The first *carbon nanotubes* that lijima observed back in 1991 were so-called *multiwalled tubes*: each contained a number of hollow cylinders of carbon atoms nested inside one another like Russian dolls. Two years later lijima and Donald Bethune of IBM independently created single-walled nanotubes that were made of just one layer of carbon atoms. Both kinds of tubes are made in similar ways, and they have many similar properties—the most obvious being that they are exceedingly narrow and long. The single-walled variety, for example, is about one nanometer in diameter but can run thousands of nanometers in length.

What makes these tubes so stable is the strength with which carbon atoms bond to one another, which is also what makes diamond so hard. In diamond the carbon atoms link into four-sided tetrahedra, but in nanotubes the atoms arrange themselves in hexagonal rings like chicken wire. One sees the same pattern in graphite, and in fact a nanotube looks like a sheet (or several stacked sheets) of graphite rolled into a seamless cylinder. It is not known for certain how the atoms actually condense into tubes, but it appears that they may grow by adding atoms to their ends, much as a knitter adds stitches to a sweater sleeve.

However they form, the composition and geometry of *carbon nanotubes* engender a unique electronic complexity. That is in part simply the result of size, because quantum physics governs at the nanometer scale. But graphite itself is a very unusual material. Whereas most electrical conductors can be classified as either metals or semiconductors, graphite is one of the rare materials known as a *semimetal*, delicately balanced in the transitional zone between the two. By combining graphite's semimetallic properties with the quantum rules of energy levels and electron waves, carbon nanotubes emerge as truly exotic conductors.

For example, one rule of the quantum world is that electrons behave like waves as well as particles, and electron waves can reinforce or cancel one another. As a consequence, an electron spreading around a nanotube's circumference can completely cancel itself out; thus, only electrons with just the right wavelength remain. Out of all the possible *electron wavelengths*, or *quantum states*, available in a flat graphite sheet, only a tiny subset is allowed when we roll that sheet into a nanotube. That subset depends on the circumference of the nonatube, as well as whether the nanotube twists like a barbershop pole.

Slicing a few electron states from a simple metal or semiconductor won't produce many surprises, but semimetals are much more sensitive materials, and that is where carbon nanotubes become interesting. In a graphite sheet, one particular electron state (which physicists call the *Fermi point*) gives graphite almost all of its conductivity; none of the electrons in other states are free to move about. Only one third of all carbon nanotubes combine the right diameter and degree of twist to include this special Fermi point in their subset of allowed states. These nanotubes are truly metallic nanowires.

The remaining two thirds of nanotubes are *semiconductors*. That means that, like silicon, they do not pass current easily without an additional boost of energy. A burst of light or a voltage can knock electrons from valence states into conducting states where they can move about freely. The amount of energy needed depends on the separation between the two levels and is the so-called *band gap* of a semiconductor. It is semiconductors' band gaps that make them so useful in circuits, and by having a library of materials with different band gaps, engineers have been able to produce the vast array of electronic devices available today.

Carbon nanotubes don't all have the same hand gap, because for every circumference there is a unique set of allowed valences and conduction states. The smallest-diameter nanotubes have very few states that are spaced far apart in energy. As nanotube diameters increase, more and more states are allowed and the spacing between them shrinks. In this way, different-size nanotubes can have band gaps as low as zero (like a metal), as high as the band gap of silicon, and almost anywhere in between. No other known material can be so easily tuned. Unfortunately, the growth of nanotubes currently gives a jumble of different geometries, and researchers are seeking improvements so that specific types of nanotubes can be guaranteed.

Fat multiwalled nanotubes may have even more complex behaviour, because each layer in the tube has a slightly different geometry. If we could tailor their composition individually, we might one day make multiwalled tubes that are self-insulating or that carry multiple signals at once, like *nanoscopic coaxial cables*. Our understanding and control of nanotube growth still falls far short of these goals, but by incorporating nanotubes into working circuits, we have at least begun to unravel their basic properties.

Several research groups, including our own, have successfully built working electronic devices out of carbon nanotubes. Our *field-effect transistors* (FETs) use single semiconducting nanotubes between two metal electrodes as a channel through which electrons flow. The current flowing in this channel can be switched on or off by applying voltages to a nearby third electrode. The nanotube-based devices operate at room temperature with electrical characteristics remarkably similar to off-the-shelf silicon devices. We and others have found, for example, that the gate electrode can change the conductivity of the nanotube channel in an FET by a factor of one million or more, comparable to silicon FETs. Because of its tiny size, however, the nanotube FET should switch reliably using much less power than a silicon-based device. Theorists predict that a truly nanoscale switch could run at clock speeds of one terahertz or more—1,000 times as fast as processors available today.

The fact that nanotubes come with a variety of hand gaps and conductivities raises many intriguing possibilities for additional nanodevices. For example, our team and others have recently measured joined metallic and semiconducting nanotubes and shown that such junctions behave as diodes, permitting electricity to flow in only one direction. Theoretically, combinations of nanotubes with different band gaps could behave like light-emitting diodes and perhaps even *nanoscopic lasers*. It is now feasible to build a *nanocircuit* that has wires, switches and memory elements made entirely from nanotubes and other molecules. This kind of engineering on a molecular scale may eventually yield not only tiny versions of conventional devices but also new ones that exploit quantum effects.

We should emphasize, however, that so far our circuits have all been made one at a time and with great effort. The exact recipe for attaching a nanotube to metal electrodes varies among different research groups, but it requires combining traditional lithography for the electrodes and higher-resolution tools such as atomic force microscopes to locate and even position the nanotubes. This is obviously a long way from the massively parallel, complex and automated production of microchips from silicon on which the computer industry is built. Before we can think about making more complex, *nanotube-based circuitry*, we must find ways to grow the nanotubes in specific locations, orientations, shapes and sizes. Scientists at Stanford University and elsewhere have demonstrated that by placing spots of nickel, iron or some other catalyst on a substrate, they can get nanotubes to grow where they want. A group at Harvard University has found a way to merge nanotubes with silicon nanowires, thus making connections to circuits fabricated by conventional means.

These are small steps, hut already they raise the possibility of using *carbon nanotubes* as both the transistors and the interconnecting wires in *microchip circuits*. Such wires are currently about 250 nanometers in width and are made of metal. Engineers would like to make them much smaller, because then they could pack more devices into the same area. Two major problems have so far thwarted attempts to shrink metal wires further. First, there is as yet no good way to remove the heat produced by the devices, so packing them in more tightly will only lead to rapid overheating. Second, as metal wires get smaller, the gust of electrons moving through them becomes strong enough to bump the metal atoms around, and before long the wires fail like blown fuses. In theory, nanotubes could solve both these problems. Scientists have predicted that carbon nanotubes would conduct heat nearly as well as diamond or sapphire, and preliminary experiments seem to confirm their prediction. So nanotubes could efficiently cool very dense arrays of devices. And because the bonds among carbon atoms are so much stronger than those in any metal, nanotubes can transport terrific amounts of electric current—the latest measurements show that a bundle of nanotubes one square centimeter in cross section could conduct about one billion amps. Such high currents would vapourize copper or gold.

Carbon nanotubes have a second interesting electronic behaviour that engineers are now putting to use. In 1995 a research group at Rice University showed that when stood on end and electric field, carbon nanotubes will act just as lightning rods do, concentrating the electrical field at their tips. But whereas a lightning rod conducts an arc to the ground, a nanotube emits electrons from its tip at a prodigious rate. Because they are so sharp, the nanotubes emit electrons at lower voltages than electrodes made from most other materials, and their strong carbon bonds allow nanotubes to operate for longer periods without damage.

Field emission, as this behaviour is called, has long been seen as a potential multibillion-dollar technology for replacing bulky, inefficient *televisions* and *computer monitors* with equally bright hut thinner and more power-efficient *flat-panel displays*. But the idea has always stumbled over the delicacy of existing field emitters. The hope is that nanotubes may at last remove this impediment and clear the way for an alternative to *cathode-ray tubes* and *liquid-crystal panels*.

It is surprisingly easy to make a high-current field emitter from nanotubes: just mix them into a composite paste with plastics, smear them onto an electrode, and apply voltage. Invariably some of the nanotubes in the layer will point toward the opposite electrode and will emit electrons. Groups at the Georgia Institute of Technology, Stanford and elsewhere have already found ways to grow clusters of upright nanotubes in neat little grids. At optimum density, such clusters can emit more than one amp per square centimeter, which is more than sufficient to light up the phosphors on a screen and is even powerful enough to drive microwave relays and high-frequency switches in cellular base stations.

Indeed, two companies have announced that they are developing products that use carbon nanotubes as field emitters. Ise Electronics in Ise, Japan, has used nanotube composites to make prototype vacuum-tube lamps in six colours that are twice as bright as conventional light bulbs, longer-lived and at least 10 times more energy-efficient. The first prototype has run for well over 10,000 hours and has yet to fail. Engineers at Samsung in Seoul spread nanotubes in a thin film over control electronics and then put *phosphor-coated glass* on top to make a prototype flat-panel display.

The third realm in which carbon nanotubes show special electronic properties is that of the very small, where size-dependent effects become important. At small enough scales, our simple concepts of wires with resistance dramatically fail and must be replaced with quantum-mechanical models. This is a realm that silicon technology is unlikely to reach, one that may yield surprising new discoveries but will also require significantly more scientific research than will either nano circuits or nanotube field-emission devices.

For example, researchers are currently debating exactly how electrons move along a nanotube. It appears that in defect-free nanotubes, electrons travel "*ballistically*"—that is, without any of the scattering that gives metal wires their resistance. When electrons can travel long distances without scattering, they maintain their quantum states, which is the key to observing effects such as the interference between electron waves. A lack of scattering may also help explain why nanotubes appear to preserve the "*spin*" state of electrons as they surf along. (*Electron spin* is a quantum property, not a rotation.) Some researchers are now trying to make use of this unusual behaviour to construct "spintronic" devices that switch on or off in response to electrons' spin, rather than merely to their charge, as electronic devices do.

Similarly, at the small size of a nanotube, the flow of electrons can be controlled with almost perfect precision. Scientists have recently demonstrated in nanotubes a phenomenon called *Coulomb blockade*, in which electrons strongly repulse attempts to insert more than one electron at a time onto a nanotube. This phenomenon may make it easier to build *single-electron transistors*, the ultimate in sensitive electronics. The same measurements, however, also highlight unanswered questions in physics today. When confined to such skinny, one-dimensional wires, electrons behave so strangely that they hardly seem like electrons anymore.

Thus, in time, nanotubes may yield not only smaller and better versions of existing devices but also completely novel ones that wholly depend on quantum effects, Of course, we will have to learn much more about these properties of nanotubes before we can rely on them. Some problems are already evident. We know that all molecular devices, nanotubes included, are highly susceptible to the noise caused by electrical, thermal and chemical fluctuations. Our experiments have also shown that contaminants (oxygen, for example) attaching to a nanotube can affect its electrical properties. That may be useful for creating exquisitely sensitive chemical detectors, but it is an obstacle to making single-molecule circuits. It is a major challenge to control contamination when single molecules can make a difference.

Nevertheless, with so many avenues of development under way, it seems clear that it is no longer a question of whether nanotubes will become useful components of the electronic machines of the future but merely a question of how and when.

14

NANO TOOLS

Nanotechnology is not a set of particular techniques, devices, or products. It is, rather, the set of capabilities that we will have when our technology gets near the limits set by atomic physics. We can make predictions for such a technology without knowing the specifics of how it will be achieved. We can, for example, know the strength of a substance with a given pattern of atoms and covalent bonds without knowing the process by which it was formed. We do have to know however, the pattern of atoms and bonds. The laws of physics don't tell us directly how strong a material can be; they tell us how strong a particular one will be. It is similar with other things of interest to a technologist. Physical law does not tell us how powerful a motor can be; it lets us say how powerful a specific motor will be. So we can only get a grasp of the outlines of the capabilities of nanotechnology by analyzing a set of designs. So we have to come up with a set of designs.

These designs have to be simple enough so that we can analyze them now. The more they look and act like machines we already know about, the less predicting we have to do and the more likely we are to get it right. Note that this is not necessarily the way many actual *nanomachines* will be built, and almost certainly not the most efficient, powerful, or economical way to build them. But if we can design nanomachines like current-day macroscopic ones, we know all the things that need to be analyzed, from structural strength to frictional heating, and we will have good reason to believe that they will work. That means that the real nanomachines of future designs will almost certainly be better than our designs here. As we know for sure that they will be at least as good.

Another property that we would like to have in our future technology is *autogeny*. An autogenous technology is one whose manufacturing base—all the machines that make machines—is capable of producing any piece of machinery in the manufacturing base. In our context, it means designs that would, if they existed, be able to build more machines like themselves. An example is today's machine tool industry: it consists of machines of steel cut into shapes with a ten-thousandth of an inch accuracy, and it can make machines of steel cut into shapes with a ten-thousandth of an inch accuracy. It's a stable point in the panorama of machine-building possibilities.

Scaling Laws

An ant does not look like an elephant. An ant can lift hundreds of times its own weight, but an ant the size of an elephant could not even stand up. The reason is *scaling laws*. The one directly applicable to the ant/elephant question is called the *square/cube law*. The weight of an object varies with its volume, that is, with the cube of its length, where the strength of its legs varies with its cross-sectional area, and thus the square of the length. In other words, an ant the size of an elephant, one thousand times as long, would have legs a million times stronger, but it would weigh a billion times as much.

If we run in the opposite direction, down to *nanomachines*, it turns out we can just about ignore weight entirely.

Do not try to bend a windowpane of glass in your hands. It would not bend enough for you to notice, and if it breaks you will cut yourself. But an optic fiber and the fibers in the fiberglass fabric of an auto/boat patch kit, bend easily. You could tie knots too small to see in the ultrafine fibers of fiberglass insulation. They are all the same stuff: glass. Being thinner means being more flexible—another scaling law. Parts in nanomachines are almost inconceivably thin, and so we will have to use the stiffest materials we can, like diamond, to compensate.

Swing your arm in a full circle, taking one second to do so. Your hand will move at between one and two meters per second. A robot arm a million times shorter than yours would make a circle a million times shorter, so if it moved at the same speed, it'd go around a million times per second.

Of course, if you tried to swing your arm around a million times per second, it would fly right off. Your bones and tendons are way too weak to support that kind of force. But the scaling laws say that the tiny robot arm will experience the same stress per unit area (in general) as yours when its hand moves at the same speed. So its frequency can go up as its size goes down. You can see the same phenomenon at work in the animal kingdom. Pelicans and geese flap their wings no faster than you can wave your arms. Sparrows and wrens flap much faster, and a hummingbird's wings are a blur.

You could pick up and place, with reasonable accuracy, small items from one spot on your desk to another at about one per second, but a nanoscale robot arm should be able to do the same at about a million second. While car engines run at thousands of *revolutions per minute* (RPM) and small electric motors at tens of thousands, nanoengines and motors can turn at billions or tens of billions of RPM, well within conservative design parameters.

Structures

Today's macroscopic machinery is mostly made of metal, although plastic of various kinds has played an increasing role for the past half-century. *Steel* is king, with *aluminum* alloys playing the role of prince. But we cannot make *nanomachines* from steel, or aluminum or plastic either. It is those *scaling laws*.

The same mathematics of form that make fiberglass flexible means that a beam of steel at the nanoscale would be about as useful as one of Play-Doh in the everyday world. To make useful beams, frameworks, casings, and shafts, we need to use the stiffest, hardest stuff we have: diamond. (Some theorists have claimed that a structure of carbon nitride is harder, but no one's managed to make it yet.) Even diamond is a bit rubbery at the nanoscale. It would be nice to have something harder, but diamond will do.

As I write, it is diamond's golden anniversary: synthetic diamond, that is. Just fifty years ago, diamond, real honest-to-goodness diamond, of the same molecular structure as natural gems, was first made in the laboratory. Today, it's made by the ton in factories: about a billion dollars' worth of synthetic industrial diamond is sold each year.

Nature produces diamond by compressing carbon at extreme temperatures and pressures. The first industrial method for *diamond synthesis* worked much the same way, producing spoonfuls of diamond grit in gargantuan one-hundred-ton hydraulic presses. But the way most diamond is made today is by chemical vapour deposition, or CVD. This is a method whereby, at relatively low temperatures and pressures, one can grow diamond crystals from carbon-bearing gas such as methane (natural gas) or acetylene. Even alcohol has been used.

Some reactions are very finicky. CVD is not. Robert deVries, retired General Electric diamond researcher, *quipped*, "You can almost make this at home with a *microwave oven.*" The process has been duplicated by amateurs in their garages. In any chemical process, you can control the conditions only on the average. The actual reactions involve random collisions between the gas molecules and the diamond being formed. But in a nanomachine, everything is precisely controlled. You can specify just which molecule touches which, when, and how hard. Robert Freitas and Ralph Merkle have made an extensive study of the process. Given how easy diamond making is with random gas collisions, it is unimaginable that it would not be quite straight forward in a nanomachine.

Diamond is a wonderful building material. It's five times as hard and twenty times as strong as steel. (That's high-alloy steel. It's one hundred times as strong as the "*mild steel*" that coat hangers are made of.) It is as slippery as Teflon. It is a good conductor of heat and a good insulator of electricity, both excellent qualities in a framework material. And its thermal expansion is tiny. So we'll be using a lot of diamond (and other gemstone or gemlike materials: sapphire, tungsten carbide, boron nitride, etc.) in our *nanomachines*. Cylindrical chunks of diamond work fine for shafts, and slabs of it work fine for walls and supports.

Bearings

The most important element of today machine technology is the bearing. The simplest thing a part can do repeatedly is turn, and *bearings* allow it to do this while supporting a force. In your car, the crank shaft, driveshaft, axles, and wheels all turn, supported by bearings. If the bearings for the power of the engine is converted by friction into heat instead of motion, and the car seizes up or catches fire. The simplest form of a bearing is a *sleeve bearing*, simply a shaft going through a hole. The harder the material of the shaft and hole, the better; that why jeweled bearings are used in, fine watches. A door hinge is a common example of a sleeve bearing. Like any macroscopic bearing, a hinge turns ignore easily with a few drops of oil on it.

At the *nanoscale*, a molecule of oil is a lot more like a loose part than a lubricant. On the other hand, we can do something unavailable at the macroscale, which is to make tile bearing surfaces atomically precise. That as smooth as the laws of physics allow. They are still "*bumpy*" with individual atoms, but atoms are soft and slippery. A common lubricant at the macroscale is graphite. Graphite is a form of carbon that comes in stacks of sheets, where each sheet is a single molecule, atomically smooth. The sheets slide over each other easily. At the *nanoscale*, we can roll the sheets up into seamless tubes. Two such tubes, nested inside each other, should you lake a very nice bearing.

They do such tubes, called *buckytubes* or *fullerene nanotubes*, can be made now, on a hit-or miss basis, in enough quantity to be experimented on in the laboratory. Two of them nested one inside the other are sometimes called a *double-walled nanotube* (DWNT). Shell nanotube bearings demonstrated in the lab, with the outer tube rotating on the inner and carrying a load. I am sure that lots of graphite hearings will be used in nanomachines. Forever, they have a couple of drawbacks. First, they have no resistance to an axial force: the tubes are just as happy to slide in a linear fashion as to rotate. For some applications, this is just what you want. Consider the brake cables on a bicycle. For others, it is not. You do not want the axles on your car sliding back and forth.

The other problem is that graphite is soft. In many applications in a machine, you want the bearings, and everything else, to be as hard and stiff as possible, If your door hinges were made of rubber, the door would sag, scrape the floor, and hit the jamb. The stiffer the bearings (and the parts), the more likely the parts will go where they were planned to. We can design (but not yet build at the nanoscale) bearings of diamond and similar hard crystals, though jeweled bearings have been used in fine watches for centuries. With careful attention to the configuration of surface atoms, these can be quite slippery to rotation, but resist axial and radial forces letter than graphite ones.

Macroscopic sleeve bearings are cheap and ubiquitous, but for more demanding applications, more sophisticated forms of bearings are used. One major reason is friction. There is no friction at the atomic scale, but there are effects that cause some of the same problems. At the macroscale, friction causes wear. A properly designed nanomachine will not wear. A wear occurs because the surface imperfections of macroscopic parts, or grit, dislodge material when surfaces move in contact. A *nanomachine*, if properly designed and built, has no surface imperfections and no grit. (This does not mean that a nanomachine will last forever. One good hit from a cosmic ray and it's a pile of junk.)

Heating, the other effect of friction, is something we do have to worry about. *Atomically* smooth doesn't mean mathematically smooth.

So when two surfaces made of atoms move past each other, the atoms of one will press alternately against the atoms of the other and the gaps in between. Now, there is no dragging force as one atom slides on another; atoms themselves are perfectly smooth. But an atom moving across a bumpy surface gets bounced up and down, and that bouncing is heat. Rub your fists together, with your knuckles in contact. You'll feel the vibration all the way up your arms. That vibration represents lost energy in a nanomachine.

You can cut clown this vibration a lot by giving the two surfaces in contact a different pitch, so that one has a different number of atoms per nanometer than the other. You can get an intuitive feeling for the difference by sliding a couple of machine screws along each other. Two screws of the same pitch will lock up as the grooves match; screws of different pitch will slide. Even so, there is still some energy dissipation from sliding surfaces.

At the macroscale, more sophisticated bearings use balls or rollers. Surfaces do not slide, they roll. The same techniques work fine at the nanoscale. You have to remember that even diamond is rubbery at the nanoscale; rolling across it is like rolling across a box spring. You can lose energy by making the whole thing jiggle. But that can be minimized as well, especially if the bearing is designed to run at one specific speed. The natural vibration of the part can be tuned to feed the energy back into the mechanical motion rather than dissipating it as heat.

The next machine element we need is the gear. Here the bumpiness of atoms is an advantage. It is quite possible to have gears whose teeth are atoms or rows of them. In this application, as with *macroscopic gears*, the important thing is to get; the pitch to match so they do not slide easily.

Since atoms are soft and diamond is rubbery, you can—in fact, you must—press gears together, where in a *macroscopic machine* this would cause jamming. In a *nanomachine*, however, this means that the gears are supported and fewer hearings are necessary. In fact, for some designs, there is no distinction. A planetary gear and a roller bearing can be the same thing.

Motors

The subject of fueled engines deserves its own chapter, but we'll consider electric *motors* here. *Electric motors* at the macroscale work by magnetism. Current flowing through wires creates magnetic fields that exert physical forces between the wires and other current-car crying wires or permanent magnets. Unfortunately this won't work at the *nanoscale*. It is those scaling laws again. The magnetic effect gets less power fail as the scale goes down. Luckily, the electric effect gets more powerful, so well use it instead. The electric effect is powerful enough to cause motion even at the macroscale—it's what causes static cling in clothing. At the nanoscale it is almighty powerful. In fact, it is what holds atoms and molecules themselves together.

The simplest way to design an *electrostatic motor* is much like a waterwheel: electrons are pumped in at the top, they fill buckets on the wheel, and they are sucked off at the bottom. Instead of gravity pulling them down, the electric force pushes them from a low voltage area to a high voltage one (low to high because electrons are negatively charged—blame Benjamin Franklin).

The power of a motor is its torque (rotary force) times its speed. For this kind of electric motor, ignoring friction and the like, power is also the voltage dines the current. In fact, the torque is proportional to the voltage and the speed determines the current: the faster those little buckets go, the more electrons are being carried across per second.

Drexler, in *Nanosystems*, analyzes a motor of this type. Its wheel is 390 nanometers in diameter and 25 nanometers thick. It rotates at 800 megahertz (48 billion RPM). The buckets of electrons are moving at about Mach 3 (faster than 2000 mph—the interior of the motor must maintain a vacuum). It draws 110 nanoamps of current at 10 volts, and produces 1.1 microwatts of power.

This may not seem like much, because the motor is tiny. A billion of them, producing 1.1 kilowatts, would be about the size of the ball at the point of a ballpoint pen. Enough of them to match the 100,000 horsepower of a large commercial jet engine would fit in the palm of your band.

Oh, and by the way, Drexler's design is overly bulky, since he was designing a combined motor and flywheel. The motor could be designed to work fine in at least fifteen times less volume. The bottom line is that for any macroscopic device using nanotechnology, motors will never be seen. Microscopic motors can be strewn throughout the fabric of whatever you are designing, capable of providing more physical force or power than you could ever need.

For applications where a precise position is needed instead of top speed and power, a stepping motor, similar to a *clock motor*, is indicated. Like other motors, macroscopic steppers are electromagnetic but the nano-sized ones would be electrostatic. One way to build such a motor would be to separate electric charges in the rotor (this is easy—even so common a molecule as water has a built-in charge separation) so that its has alternating areas of positive and negative charge. Then surround it with electrodes that you charge and discharge with alternating currents. Driving this kind of motor is a lot more complex than the other kind, but you can make it go for backward, stop, turn a specific angle, and so forth. Stepping motors are commonly used today in robotics.

From Here to Autogeny

Now that we have motors, shafts, gears, and bearings, we can build a robot arm. An arm is basically just a series of boxes on hinges with a motor to power each joint. Typically the motors are geared down to provide more force and higher precision (and stiffness) than the motor would have alone. Robot arms with grippers (a finger is just a miniature arm) can pick up things and move them, and in particular can build more robot arms. Macroscopic parts are made in many different ways, ranging from molding, to forging (beat it into shape), to machining. Machining is the most expensive, often the slowest method, but also the most accurate and least restricted in what it can make. A milling machine and a lathe can make just about any machine part (you need special attachments for screws and gears).

Milling machines and *lathes* can he thought of as very specialized robot arms. What they do is hold a knife and carve the shape you want out of the chunk of material you give them. To get the shape just right, they must hold the knife very firmly. This is technically known as *stiffness*, and it tells you how far the knife (or whatever) will move given a certain force against it. The higher the stiffness, the better.

A milling machine doesn't look like an arm. All the joints are big, heavy, precision-machined hunks of metal that don't wobble at all. If we build a nanomachine on the plan of a milling machine, the position of the tool could be have exactly controlled with respect to the position of the work piece. And this is the kind of machine that we'll make to build the parts that the slender robot arms will put together. There is just one big difference. A classic milling machine (or lathe) operates by taking a chunk of material and cutting off every thing that does not look like the part you need. At the nanoscale, a better strategy would be the opposite: to build up the part you want, an atom or molecule at a time. This is what allows you to have complete control over where every atom is in the finished product.

This method of controlling the reaction precisely is called *positional chemistry* or *mechanosynthesis*. Mechanosynthesis is what will allow us to make million—atom parts and get every atom exactly right. This, in turn, will allow us to build molecular milling machines and do mechanosynthesis. And that is our *autogenous technology*.

Pumps

Unless you live on an old-fashioned farm and milk your cows by hand, the milk you drink came through a milking machine and subsequent processing and handling stages, courtesy of a number of *pumps*. Unless you own an artesian well, the water you drink was pumped, either from the ground or by way of water mains from a reservoir. Commonly used fuels are pumped, many times, through processing plants, from one tank to another along pipelines, into your car at the service station, and again from your gas tank to your car's engine.

Pumps for air in low-pressure applications are called *fans*. There are probably fans in your computer to cool it, fans in your home heating system to heat and cool you, and more fans in your car. A typical hydronic home heating system has both fans and pumps.

Moving fluids around and managing their pressure is a substantial application of, and a necessity for, modern technology. Nanomachines will need to do it, too. If anything, most of the designs for dumps at the macroscale will work better at the *nanoscale*. The simplest pump is fan. Fans are noisy and inefficient because they cause turbulence. At the nanoscale they won't, because the smaller you get, the more syrupy fluids act. For gases, fans can be made that interact more or less individually with the molecules. Molecules of a gas bounce off one another like billiard balls, and fly some distance before they hit another molecule. The average distance for a given gas at a given pressure is called the mean free path. If the length of a molecule's path through the fan is smaller than the mean free path, molecules going through the fan tend to interact with the fan and not one another. If the fan blades can move for than the molecules tend to, the fan can attain something like a compression ratio of ten with one rotor and one stator (or pair of counter-rotating fans). Ordinary air has a mean free path of about 60 nanometers and an average speed in the neighbourhood of the speed of sound—figures easy for a *nanofan* to match.

One common pump for liquids, the *centrifugal pump*, is not recommended at the nanoscale for scaling-law reasons. But the kind you probably think of when someone says *pump*, the piston or positive displacement pump, should work fine. The positive displacement pump usually consists of two one-way valves and a cylinder in which a piston goes up and down. At the macroscale, the valves are often spring valves, but at the nanoscale it will probably be preferable to make them mechanically actuated in a timed sequence with the piston, like the valves in a car engine.

Can we make a *positive displacement pump* that interacts with the fluid one molecule at a time? Yes, and in a fairly elegant design. Consider a cylinder with a set of molecule-shaped dents in it. It is embedded in a wall like a revolving door. The molecules go through as the cylinder turns, just as people go through the door. But the trick is that a little piston at the bottom of each dent pushes the molecule out when it on the output side, and retracts to let a new molecule in on the input side.

Doing it this way allows you to make the dents much more attractive to one kind of molecule than to another, by making them fit snugly. This is how a large proportion of the molecular machinery in your body works—it is how *hemoglobin* grabs oxygen molecules to carry them in your red blood cells, for example. With a set of dents that strongly favours one kind of molecule, we can pump that fluid preferentially out of a solution—for example, oxygen out of air, fresh water out of seawater. This kind of selective pump is called a *molecular sorting rotor*.

Nanocircuitry

Light and radio are waves in the same stuff: electromagnetic fields. The only difference is the length of the waves, and thus the frequency. With fast enough *nanoswitches* and appropriated sized antennas, we should be able to handle light with the same facility that we now handle radio waves. Today, for example, we have phased-array antennas for radar. A flat array of antennas can produce a beam going in any desired direction by having each little antenna match the phase properly. What is more, it can detect the direction of an incoming beam without a lens or mirror. In essence, it a controllable hologram at microwave frequencies.

With *nanoelectronics nanoantennas*, we should be able to do the same thing with light—phased-array optics. Cover the surface of an object with optical antennas and it should be able to take on any properties you like that involve the reflection and absorption of light, or generate a completely synthetic pattern, making it appear to be anything you want. You could even have it present the image that would be there even in its absence, making it effectively invisible.

In order to make antennas and to connect switches, you need wires. Macroscopic wires are generally made of metal. Metal conducts electricity because the electron (some of them anyway) are not nailed down to particular nuclear or in particular bonds, as they are in diamond, for example, but are free to roam around at the behest of any passing electric field. Graphite conducts electricity because electrons can roam, but only within a given sheet. Rolled-up graphite, or buckytubes, can conduct or not depending on their particular structure. Various molecules have areas in which electrons can move around. These areas act as optical antennas at various frequencies, which is why different substances have different colours and albedos.

Conductive paths at the *nanoscale* can be made in many ways besides laying down wires of metal, but wires of metal do work. The other ways just: give us other options. Some arrangements of atoms form *superconductors—substances* with no electrical resistance at all. Since superconductors can be made with the statistical mixing of current-day chemistry it seems likely that we could make better ones with a technology that allowed us to place atoms in a specific, desired pattern. Superconducting wires would be useful at the nanoscale, since ordinary wires gain more resistance as they get thinner. It remains to be seen whether superconducting wires can be made thin enough to be useful, however.

Putting it all Together

Most of the elements—parts and mechanisms—of macroscopic machines can be scaled to work at molecular size, like sleeve bearings, or redesigned to, like electric motors. The few that cannot, like *centrifugal pumps*, have substitutes, plus there are many new designs, like sorting rotors, that don't have any macroscopic parallel at all. At we can confidently expect to be able to design and build machine systems at the *nanoscale* with a wide variety of capabilities and applications, and in particular, to design and build manufacturing systems.

15

ATOMIC FORCE MICROSCOPY

The *atomic force microscope* (*AFM*) is one of the many recently developed types of nanometer-resolution microscopes or *scanning probe microscopes* (*SPM*). The AFM has a much wider range of applicability than the other SPMs: it can be used to observe and manipulate nanometer-sized objects of both conductive and insulating nature in both vacuum, air, gasseous, and liquid environments. The following sections provide a more detailed treatment of the physical principles behind the AFM.

BASIC PRINCIPLES OF THE AFM

A sketch of the layout of a typical AFM setup is shown in Fig. 15.1. The central component is the nanometer-sized tip mounted on the elastic cantilever. By use of the piezo-electric *xyz* scan driver the

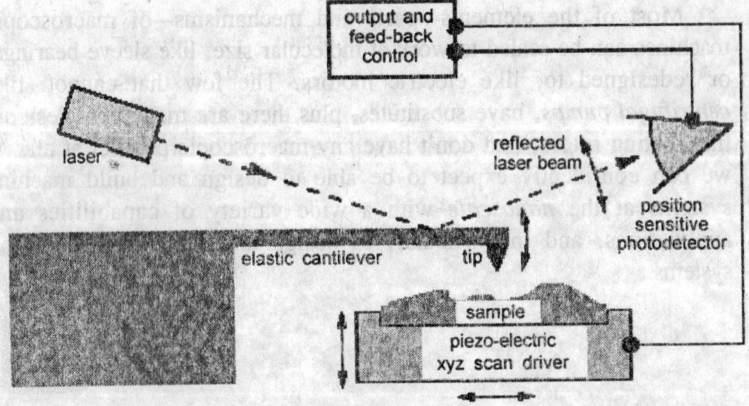

Fig. 15.1. A sketch of the layout of a typical AFM set-up showing the key components.

tip is scanned across the sample. The distance between the tip and the sample can be anywhere between 0 and 100 nm. Dependent on the topography and contents of the surface of the sample the force acting on the tip changes during the scan resulting in a position dependent deflection of the elastic cantilever. The deflection is monitored by reflecting a laser beam on the cantilever and recording the movement of the reflected beam with a position sensitive *photodetector*. The signal from the photodetector can be used in a feedback loop with the *xyz* scan driver to control the motion of the tip.

To build a well-functioning AFM several requirements has to be fulfilled:

1. The spring constant of the cantilever should be small enough to allow detection of minute atomic forces.
2. The resonance frequency of the cantilever should be as high as possible to minimize sensitivity to external mechanical vibrations.
3. The tip should be as sharp as possible to allow atomic resolution.
4. The tip should be as narrow as possible to allow penetration into deep troughs on the surface.

We shall look into these requirements in the following sections, where the different modes of operations are described.

CANTILEVER: SPRING CONSTANT AND RESONANCE FREQUENCY

The actual design geometry of an AFM cantilever can vary. For simplicity we shall in the following analysis just consider a beam of length L and rectangular cross section with height h and width w. Due to interactions with the surface atoms of the sample an external force F_{tip} acts on the tip and bends the cantilever.

It is a simple exercise of applied elasticity theory to find the bending Δz of the tip due to the force F_{tip}. Hooke's law for elastic solids relates the force per area (the stress) σ_{xx} with the relative change in length (the strain) Δ/L_0,

$$\sigma_{xx} = Y\frac{\Delta L}{L_0}, \qquad ...(1)$$

where Y is Young's modulus, a material parameter we already encountered and calculated. Due to the bending the cantilever is compressed on the lower surface, $L < L_0$ at $z = -h/2$, elongated on the upper surface, $L > L_0$ at $z = +h/2$, and unchanged at the center plane, $L = L_0$ at $z = 0$. Thus the length $L(z)$ of the cantilever is a function of z. We assume that the bended cantilever has the circular geometry depicted in Fig. 15.2(b). This is an approximation that will

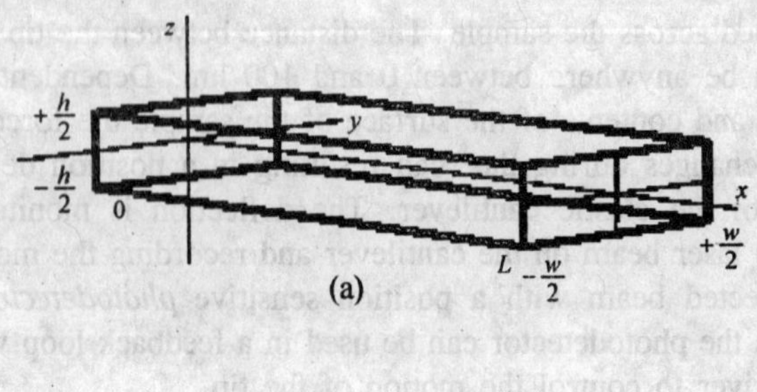

(a)

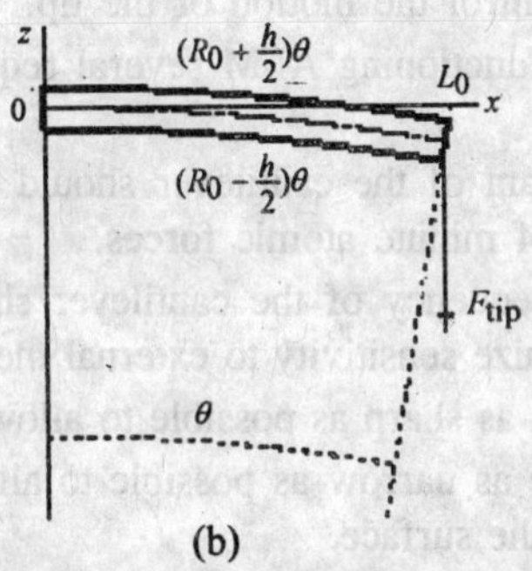

(b)

Fig. 15.2. (a) An AFM cantilever of dimensions L×w×h. (b) The approximately circular bending of the cantilever due to the external force F_{tip} acting on the tip.

allow us to obtain decent results without too much math. The math is simplified due to the fact that a circle has a constant radius of curvature. The exact results are quoted along the way.

At Fig. 15.2(b) we see that there is an angle θ between the end planes at the two ends at $x = 0$ and $x = L$. If we denote the radius of the bended but unstretched center plane $z = 0$ by R_0 we have $L_0 = R_0\theta$ and deduce that

$$L(z) = (R_0 + z)\theta = \left(1 - \frac{z}{R_0}\right)L \Rightarrow \frac{\Delta L}{L_0} = \frac{L - L_0}{L_0} = \frac{z}{R_0}. \quad ...(2)$$

The *external force* F_{tip} acts on the tip and bends the cantilever. However, even in this situation the cantilever is in equilibrium, and consequently the external *force torque* L_0F_{tip} around the line $(x, z) = (0, 0)$ at the left end of the cantilever must balance the internal torque set up by the internal stress forces σ_{xx} times the arm z:

$$L_0F_{tip} = \int_{-\frac{w}{2}}^{+\frac{w}{2}} dy \int_{-\frac{h}{2}}^{+\frac{h}{2}} dz\, z\, \sigma_{xx} = w\int_{-\frac{h}{2}}^{+\frac{h}{2}} dz\, z\, Y\frac{z}{R_0} = \frac{1}{12}Y\frac{wh^3}{R_0}$$

$$= \frac{1}{6}Y\frac{wh^3}{L_0^2}\Delta z, \quad ...(3)$$

where Δz is the vertical displacement of the cantilever at $x = L$. It is given by $\Delta z = R_0(1 - \cos\theta) \approx \frac{1}{2}R_0\theta^2 = \frac{L_0^2}{2R_0}$ or $1/R_0 = 2\Delta/L_0^2$. So the force F_{tip} can be expressed by the displacement Δz of the tip and a spring constant K:

$$F_{tip} = K\Delta z, \qquad \text{...(4a)}$$

$$K = \frac{1}{6}\left(\frac{h}{L}\right)^3 wY \text{ (the circular shape approximation),} \qquad \text{...(4b)}$$

$$K = \frac{1}{4}\left(\frac{h}{L}\right)^3 wY \text{ (the exact result).} \qquad \text{...(4c)}$$

From the spring constant we can estimate the (smallest) resonance frequency ω_0 of the cantilever (corresponding to a quarter wavelength oscillation) when it is oscillating freely. It is given by $\omega_0 = \sqrt{K/m_{eff}}$. Note that it is not the cantilever mass m that appears in the formula but an effective mass m_{eff}. This is due to the fact that the various parts of the cantilever do not oscillate with the same amplitude. For $x = 0$ the amplitude is zero, while for $x = L$ it is the maximal Δz. It has already been shown how the circular shape approximation leads to $m_{eff} = m/5$ and thus

$$\omega_0^2 = \frac{K}{m_{eff}} = \frac{5}{6}\frac{Yw}{m}\left(\frac{h}{L}\right)^3 = \frac{5}{6}\frac{Yw}{\rho whL}\left(\frac{h}{L}\right)^3 = \frac{5}{6}\frac{Y}{\rho}\frac{h^2}{L^4}, \qquad \text{...(5)}$$

where we have introduced the mass density ρ. The approximate and exact results are

$$\omega_0 = 0.91\sqrt{\frac{Y}{\rho}}\frac{h}{L^2} \text{ (the circular shape approximation),} \qquad \text{...(6a)}$$

$$\omega_0 = 1.02\sqrt{\frac{Y}{\rho}}\frac{h}{L^2} \text{ (the exact result).} \qquad \text{...(6b)}$$

Eqs. (4c) and (6b) are central to the design of an AFM. The ratio h/L should be small to give a small spring constant and hence a good force sensitivity, but h/L^2 should be large enough to ensure a high resonance frequency that minimizes the sensitivity to external mechanical vibrations. The solution is to make small cantilevers. One good method is to make them in silicon since one then can take advantage of standard silicon processing techniques perfected by the microelectronics industry.

A typical choise is $L = 100\ \mu$m, $w = 10\ \mu$m, and $h = 1\ \mu$m. A deflection $\Delta z \approx 1$Å is easily detected and since for silicon $Y = 1.6 \times 10^{11}$ Pa and $\rho = 2.33 \times 10^3$ kg m^{-3}, we obtain the following spring constant K, tip-force F_{tip}, and resonance frequency ω_0 from Eqs. (4a), (4b), and (6a):

$$K = 0.40 \text{ N m}^{-1} \quad \text{...(7a)}$$

$$F_{tip} = 40 \text{ pN} \quad \text{...(7b)}$$

$$\omega_0 = 927 \text{ kHz.} \quad \text{...(7c)}$$

These numbers are good news. The typical atomic forces acting on an AFM-tip is of the order 1 nN, and mechanical noise above 1 kHz is strongly suppressed.

Contact Mode

When operating in the so-called *contact mode* the AFM tip is forced down into the surface until repulsive contact with the *core electrons* of the surface atoms is obtained. During a scan the AFM tip simply follows the topographic features of the sample with atomic resolution (if the tip is sharp enough).

To illustrate the physics of the contact mode it is sketched in Fig. 15.3(a) how the electron orbitals get deformed as two atoms approach each other. At long range the weak attractive *van der Waals force* is the dominant force. This force will be treated in the next section. At intermediate range the cloud of valence electrons surrounding the atom get in contact and become deformed as a diatomic molecule is formed. At the closest range also the core electron clouds of the two atoms touch each other and get deformed. At this point the energy increases rapidly as a function of decreasing inter-atomic distance. It is extremely costly energy wise to compress the tightly bound core electrons further.

In general it is very difficult to calculate the exact form of the energy versus distance. There are no small parameters in the problem and many atomic orbitals must be taken into account in a serious calculation. It is customary, however, to parametrize the potential by the Lennard-Jones model potential

$$E(r) = B\left[\left(\frac{A}{r}\right)^{12} - \left(\frac{A}{r}\right)^{6}\right]. \quad \text{...(8)}$$

The r^{-6}-term is well understood. The r^{-12}-term on the other hand is just an ansatz. There is no particular reason why the power of 12 has been chosen except that it is bigger than 6 and mimmicks the

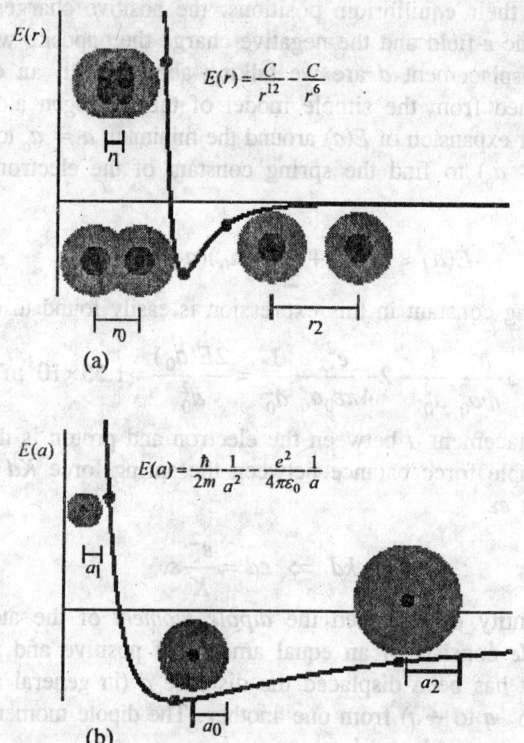

Fig. 15.3. Similarities and differences in the energy of diatomic molecules and single atoms. (b) The simple model of the ground state energy E(a) of the hydrogen atom as a function of the average distance a between the electron and the proton.

very strong repulsion observed in many experiments. The energy versus distance curve $E(r)$ ressembles the simple energy versus size-of-atom curve $E(a)$. For comparison $E(a)$ is depicted in Fig. 15.3(b).

Non-contact Mode

When operating in the *non-contact mode* the AFM-tip is not touching the surface. Instead it is influenced by the long range van der Waals forces. The advantage of the non-contact mode is that it does not destroy even the softest samples. But what is the origin of the van der Waals forces? This question will be answered now.

Atomic Polarization

The first step in understanding the van der Waals forces is a study of *atomic polarization*. When an *electric field* ε is applied to a charge neutral atom the positive and negative charges are displaced slightly from their equilibrium positions: the positive charges in the direction of the ε-field and the negative charge the opposite way. But how big a displacement d are we talking about? Well, an estimate can be obtained from the simple model of the hydrogen atom. We make a Taylor-expansion of $E(a)$ around the minimum $a = a_0$ to second order in $(a - a_0)$ to find the spring constant of the electron-proton system:

$$E(a) \approx E(a_0) + \frac{1}{2} E''(a_0)(a - a_0)^2. \qquad ...(9)$$

The spring constant in this expression is easily found to be

$$K = E''(a_0) = 3 \frac{\hbar^2}{m a_0^2} \frac{1}{a_0^2} - 2 \frac{e^2}{4\pi\varepsilon_0 a_0} \frac{1}{a_0^2} = \frac{2E(a_0)}{a_0^2} = 1.55 \times 10^3 \text{m}^{-1} \qquad ....(10)$$

The displacement d between the electron and proton is therefore found by simple force balance between the spring force Kd and the electric force $e\varepsilon$:

$$e\varepsilon = kd \;\Rightarrow\; ed = \frac{e^2}{K} \varepsilon. \qquad ...(11)$$

The quantity ed is called the *dipole moment* of the atom. An *electric dipole* consists of an equal amount of positive and negative charge q that has been displaced the distance d (in general a vector pointing from $-q$ to $+q$) from one another. The dipole moment of the dipole is just the product qd.

What is then the *electrostatic energy* $U(r)$ of the dipole qd in the *electric field* $\varepsilon(r)$? The electric field is due to some potential $V_\varepsilon(r_0)$, that is $\varepsilon(r_0) = -V_\varepsilon{}'(r_0)$. We just study the case where the dipole is parallel to the electric field, so we place $+q$ at $r_0 + d/2$ and $-q$ at $r_0 - d/2$. We then obtain the energy $U(r)$ as

$$U(r_0) = +qV_\varepsilon\left(r_0 + \frac{d}{2}\right) - qV_\varepsilon\left(r_0 - \frac{d}{2}\right)$$

$$\approx q\left(+\frac{d}{2}\right)V'_\varepsilon(r_0) - q\left(-\frac{d}{2}\right)V'_\varepsilon(r_0) = -qd\varepsilon(r_0). \qquad ...(12)$$

We have seen that an electric field can induce a dipole moment, but the converse is also true: a dipole moment induces an electric field. Consider in 1D a dipole at the center of the coordinate system, $+q$ at $d/2$ and $-q$ at $-d/2$. What is then the *electric potential* $V(r)$ and the corresponding electric field $\varepsilon(r)$? Due to the opposite charges we can write their respective potentials as $+V_0(r - \frac{d}{2})$ and $-V_0(r - \frac{d}{2})$, where $V_0(x) = q/(4\pi\varepsilon_0 x)$ is the usual *Coulomb potential*. For $V(r)$ we thus get

$$V(r) = V_0\left(r - \frac{d}{2}\right) - V_0\left(r + \frac{d}{2}\right)$$

$$\approx -V_0'(r)\frac{d}{2} - V_0'(r)\frac{d}{2} = -dV_0'(r) = \frac{dq}{4\pi\varepsilon_0}\frac{1}{r^2}. \quad ...(13)$$

The electric field is just the derivative of the potential, so

$$\varepsilon(r) = -V'(r) = \frac{2dq}{4\pi\varepsilon_0}\frac{1}{r^3}. \quad ...(14)$$

We see here the characteristic r^{-3}-dependence of the electric field from a dipole.

van der Waals Forces

The *van der Waals forces* are now deduced as follows. If two charge neutral hydrogen atoms are placed at a distance r from one another they can gain energy by spontaneously generate dipole moments. Consider for example what happens if atom no. 1 by quantum fluctuations generates a dipole moment d_1. According to Eq. (14) this results in an electric field ε_1 at the site of atom no. 2. But Eq. (11) tells us that atom no. 2 as a consequence will generate a dipole moment ed_2,

$$ed_2 = \frac{e^2}{K}\varepsilon_1, \quad ...(15)$$

and thus, according to Eqs. (12) and (14), gain the energy U

$$U(r) = -ed_2\varepsilon_1(r) = -\frac{e^2}{K}(\varepsilon_1(r))^2 = -\frac{e^2}{K}\left(\frac{2d_1 e}{4\pi\varepsilon_0}\frac{1}{r^3}\right)^2$$

$$= -\left(\frac{e^2}{4\pi\varepsilon_0}\right)^2\frac{(2d_2)^2}{K}\frac{1}{r^6}. \quad ...(16)$$

We can make the following simple estimate for $U(r)$. First we note that the only length scale relevant for the charge displacement $2d_1$ is the size of the atom, a_0. Thus we take $2d_1 \approx a_0$. With the usual abbreviation $E_0 = E(a_0)$ we have $2E_0 = e^2/(4\pi\varepsilon_0 a_0)$, and finally use Eq. (10) and write $K = 2E_0/(a_0^2)$. Thus we arrive at

$$U(r) \approx -(2E_0 a_0)^2 \frac{(a_0)^2}{2E_0/(a_0^2)}\frac{1}{r^6} = -2E_0\left(\frac{a_0}{r}\right)^6. \quad ...(17)$$

Since force is the derivate of potential energy we obtain the following expression for the van der Waals force $F_{vdW}(r)$:

$$F_{vdW}(r) = -\frac{dU(r)}{dr} = 12\frac{E_0}{a_0}\left(\frac{a_0}{r}\right)^7. \qquad ...(18)$$

To see a typical number we insert an atomic diameter $r = 2a_0$ and find

$$F_{vdW}(2a_0) = 5.8 \text{ nN}. \qquad ...(19)$$

Tapping Mode

The *tapping mode* combines the good sides of the contact and the *non-contact mode* of AFM-operation. The *AFM-cantilever* is forced to oscillate with an amplitude of the order 50 nm in such a way that the tip gently touches the sample once during each cycle. The benefits are threefold. First, the tapping mode has nearly the same resolution as the contact mode. Second, the tapping mode is almost as gentle, so it can be used for soft samples. Third, the tapping mode is in fact a simple and roboust way of operating the AFM.

16

Scanning Tunneling Microscopy (STM)

Although treated first in these lecture notes, the AFM was actually preceded by the *scanning tunneling microscope* (STM). Invented in 1981 the STM was the first of many scanning probe microscopes (SPM). In 1986 the inventors Gert Binnig and Heinrich Rohrer from the IBM Research Laboratory in Ruschlikon (Switzerland), were awarded the Nobel Prize in Physics. Nowadays, SPMs can be found in many academic and industrial physics, chemistry and biology laboratories, and as we have already seen, they are used both as standard analysis tools and as high-level research instruments.

Basic Principle of the STM

The basic principle of the STM is sketched in Fig. 16.1. In contrast to the atomic force microscope the STM can only scan conducting surfaces. The reason is that the instrument relies on the small tunneling current, typically in the sub-nA regime, that runs between the substrate and the metal (*tungsten*) tip. Through a computer controlled feedback loop and a piezo-element this current controls the height d in the z direction over the substrate. Scans across the surface in the xy plane are controlled by a piezo-electric element. The feedback voltage controlling the height in the z direction together with the xy scan voltage is used as output.

With the STM one can achieve a resolution of around 0.1 nm. This is better than the AFM, and the reason for this is the exponential dependence of the distance d in the transmission coefficient, $T(\varepsilon) \approx$

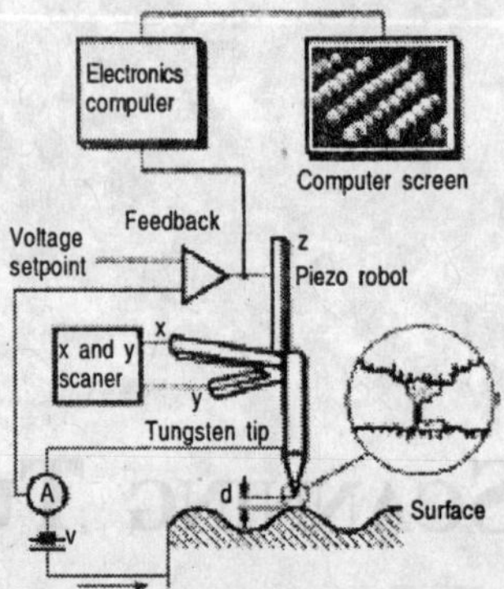

Fig. 16.1. The principle of the STM.

exp[$-4\kappa d$]. For a typical parameter values the tunneling current reduces by a factor 10 for every 0.1 nm increase in d. This means that over a typical atomic diameter of e.g. 0.3 nm, the tunneling current changes by a factor 1000. This is what makes the STM so sensitive. The tunneling current depends so strongly on the distance that it is dominated by the contribution flowing between the last atom of the tip and the nearest atom in the specimen.

The scan range for a STM is typically up to 1 μm with a scan speed of the order mm/min. The scan speed depends on the mode used for the STM.

The constant height mode is the fastest scan mode. Here the tip is kept at a fixed vertical position during the scan. The changes in the tunneling current thus reflects the electronic topology of the surface. Using this mode there is a risk that the tip will bump into unexpected high regions on the surface and be destroyed.

The constant current mode is perhaps the most widely used *STM mode*. In this mode the feedback mechanism ensures that the tunneling current is kept constant by displacing the tip vertically as the scan proceeds. This is slower than the constant height mode, but without its risk.

Piezo-electric Element and Spectroscopy

The principle of the *piezo-electric element* is sketched in Fig. 16.2. The principal effect is that for crystal lacking inverison symmetry an applied voltage in one direction makes the element longer or shorter in another direction.

A simple tripod allowing for control of the tip in all three spatial directions can be made by mounting three piezo-rods perpendicular to

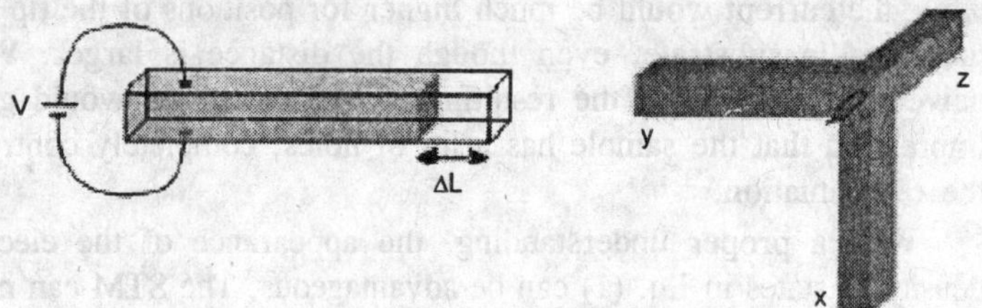

Fig. 16.2. The principle of the piezo-element. (a) Elongation of a sample piezo-electric by applying a voltage. (b) The tripod consisting of three mutually perpendicular piezo-rods.

each other. However, in most modern scanning probe microscopes, one uses a tube geometry. Each of the four indicated sections can be made longer or shorter individually. If all four sections are made longer or shorter by the same amount, the tip moves in the z direction. If the $X+$ side is made longer, and at the same time the $X-$ side is made shorter by the same amount, the tube tilts a little bit to one side, as indicated. For small deformations, this makes the tip move primarily in the x direction. The same can be done in the y direction.

Local Electronic Density of States

It is important to realize that the STM does not measure the topology but rather in some sense the electronic topology. From the discussion of quantum conductance we know that the current running between two reservoirs is given by counting how many electrons that are transferred per unit time. However, in our previous analysis we tacitly assumed that there always were states available in both reservoirs. This is true if we are dealing with bulk metals. This assumption is not generally true, and in fact we need to introduce the concept of the local electronic density of states $d(r, \varepsilon)$.

The density of state function contains the information about how many quantum states of energy ε there are available in the neighbourhood of the point r on the surface. Therefore the expression for the tunneling current $I(r)$, when the tip is placed at the position r is modified to

$$I(r) = CT(\varepsilon)\, d(r,\varepsilon) \approx A\exp[-4\kappa d]\, d(r,\varepsilon). \qquad ...(1)$$

Consequently great care must be taken when interpreting STM *scan pictures*. Imagine using the constant height mode to scan a metallic surface covered with some small non-conducting particles. Such particles have a very small electronic density of states. Therefore according to Eq. (1) the tunneling current would be very low when the tip is just above one such particle even though the distance is small, while the current would be much higher for positions of the tip above the metallic substrate, even though the distance is larger. With a naive interpretation of the resulting STM picture one would get the impression that the sample has a lot of holes, completely contrary to the real situation.

With a proper understanding, the appearance of the electronic density of states in Eq. (1) can be advantageous. The STM can namely be used as a spectrometer for measuring electronic energies. Consider a STM tip placed at a fixed position r_0. By changing the applied voltage V between the STM tip and the sample the energy ε of the electrons leaving the STM tip is changed by the amount $\varepsilon = \varepsilon_0 - eV$, where ε_0 is the energy at zero applied voltage. In an energy range where the energy dependence of the transmission coefficient is negligible $T(\varepsilon) = T$, measuring the tunneling current as a function of applied voltage V gives directly an energy scan of the density of states,

$$I(r_0, V) = CT\, d(r_0 \varepsilon - eV), \qquad \text{...(2)}$$

and thus a direct measurements of at which energies there are electronic states available in the sample. This technique has been used in numerous investigations on materials on the nanometer scale.

An Example of a STM

We end this chapter on the STM by showing an concrete example of a STM. It is taken from the Department of Physics and Astronomy, Aarhus University, Denmark. A sketch of the instrument is shown in Fig.16.3.

The setup is described as follows. The sample (1) is placed in a tantalum holder (2), which may be removed from the STM, and which is normally held down on the STM top by springs (3). The top plate is thermally and electrically insulated from the STM body by three quartz balls (10). The top plate is mounted on a 0.6 kg aluminum block which may be cooled to −160°C or heated to +130°C. The tip (4) is held by a macor holder (5), which is affixed to the top of the scanner tube (6). The scanner tube is 4 mm long with an outer/inner diameter of 3.2/2.1 mm and is mounted to the rod (7), which together with the piezo tube (9) forms a small inchworm motor used for coarse approach. The electrode of the tube is divided into three rings. In the tube two bearings are placed under the upper and the lower electrode with an extremely good fit to the rod (7).

Applying a positive voltage to an electrode will clamp that electrode to the rod whereas a negative voltage will free that electrode from

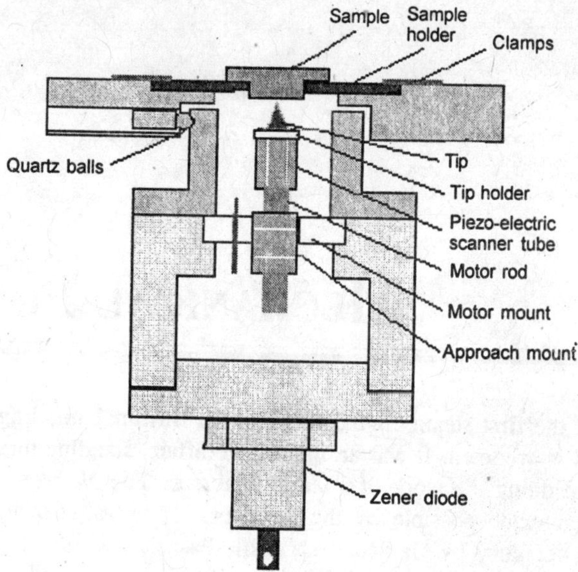

Fig. 16.3. An example of a STM.

the rod. A voltage applied to the center electrode will cause it to elongate or contract. With the right sequence of voltages applied to the three electrodes the rod will move up or down since the tube is fixed to the STM body by the macor ring (8). The motor may work in steps of down to 2Å, but at full speed it moves around 2 mm/min. The scan range is up to ± 1 um when using antisymmetrical scan voltages of ± 200 V.

The Zener diode BZY93C75 is used to counter-heat the STM body during cooling.

17

MECHANICAL POWER

In 1712 the first steam engine was built in Birmingham, England, by Thomas Newcomen. It was an impressive affair, standing three stories high and doing the work of a team of horses. The *Newcomen engine* had been made possible by the discovery of the atmosphere in the previous century by Galileo, Torricelli, Pascal, and others. It seems odd to us that the atmosphere should need to be discovered; we take it for granted. But the notion that we stand at the bottom of an ocean of air, and that the pressure it exerts on a square foot of surface is the same as the weight of a one-square-foot column of air extending to the top of the atmosphere, would have astounded Aristotle or St. Augustine.

Newcomen's engine worked directly by the action of that atmospheric pressure. The air was pushed out of a cylinder by steam. Then the steam was reverted to water by a cooling spray, leaving vacuum in the cylinder, which drew down a piston.

A horse, by a suitable arrangement of pulleys, can raise a thousand-pound weight thirty-three feet in a minute, at a pace of work that it can keep up indefinitely. Thirty horses can do in one minute the same work one can do in half an hour; in each case the quantity of work can be called a *million foot-pounds*. *Newcomen's steam engine* could do 5 million foot-pounds of work for each bushel of coal that was shoveled into it.

Newcomen's engine was a brilliant, world-changing breakthrough in technology, but it was not terribly efficient. Essentially by tinkering and optimizing the basic design, by the middle of the eighteenth century, *Newcomen engines* had been improved to produce 9 million foot—pounds of work for each bushel of coal.

Then came the Scot, James Watt. In January 1769 he patented an improved steam engine that avoided one of the big inefficiencies of Newcomen's. In *Newcomen's engine*, the working cylinder had to be cooled down to collapse the steam and produce the vacuum, and then heated up again by the new steam at each stroke. Watt introduced a separate condenser that stayed cold, and let the cylinder stay hot. Watt's engine was a major improvement, producing 28 million foot-pounds of work for each bushel of coal. The steam engine now began to be competitive economically with the water while you did not have to buy coal to run your water where, you did have to own enough length of a fairly decent-sized stream to have a ten-foot drop, and a twenty-foot drop was even better. With the steam engine you could set up shop anywhere.

One of the things that made a big difference in how much work you got out of your engine for a given amount of coal was the precision with which it was made. Better precision meant less friction, less steam escaping uselessly from joints and valves, and most of all, better reliability. From an almost entirely handmade technology in 1750, the precision and fit of metalwork improved drastically until the early 1800s, when the beginnings of a machine tool industry could be seen.

Also in the early 1800s, a new surge in steam technology happened because Watt's patents expired. Watt simply was not interested in high-pressure steam. This is probably with good reason, since the technology in which his engineering instincts were formed was not strong or precise enough to handle high pressures at high temperatures. However, by the early 1800s there had been enough advancement of fabrication and metallurgy that a new breed of engineers, notably Arthur Woolf, could successfully build engines that pushed the piston with steam pressure rather than sucking it with vacuum. This had the immediate consequence of making the engines smaller; higher pressure on a smaller piston produces the same force. What's more, they could throw away the condenser, allowing used steam to escape to the atmosphere. Woolf's double expansion pressure engine of 1804 got 56 million foot pounds of work out of each bushel of coal.

Shortly after our arrival we went to the coal wharf to see the arrival of the coal wagons which are set in motion by steam machines and which bring coals from the mines at a distance of six English miles from the wharf. It is a curious spectacle to see a number of columns of smoke winding their way through the countryside. As they approach we see them more and more distinctly until at length with a column of smoke we also perceive the wagon from which it ascends, dragging a long train of similar wagons behind it, which gives it the appearance of a monstrous serpent.

The great practical implication of smaller, more efficient engines was that you could put them on vehicles such as boats or wagons. Steam engines brought power not only for factories but transportation. The burgeoning machine tool industry provided precision parts for other machines as well, notably the spinning jenny and the power loom. In less than a century, England went from an essentially agrarian society where everything was handmade, to one with a significant manufacturing sector, using mechanical power for mass production. Railways made freight cheap and fast enough to produce an integrated national economy. This was the industrial revolution.

Meanwhile, the engineers didn't stop improving and optimizing the steam engine. By 1835 in Cornwall, the leading area of steam engineering of the day, engines could produce 125 million foot pounds of work for each bushel of coal.

No Free Lunch

Given this history in the improvement of the efficiency of steam engines, it was natural for the more philosophically inclined of engineers to wonder whether there was a limit to the process. A young French engineer named Sadi Carnot was among them, and in 1824 he published a short book entitled *Reflections on the Motive Power of Fire*. With this book, he founded the science of thermodynamics.

Carnot systematized the knowledge about heat engines first by pointing out that they work from a temperature difference, not just a source of heat: an engine sitting in an ocean of steam could not draw any power from it. This had an important implication: any heat flowing across a temperature difference in an engine was a source of inefficiency. If you used a 1,000-degree fire to heat 500-degree steam, you could, in theory, put a whole separate engine between the fire and the steam, getting extra energy out of it, and still run your original engine. So the efficient engine must transfer heat without a temperature drop. This sounds like a tall order, since we expect heat flows to be associated with temperature differences. But it can be approximated by having the working fluid and the heat source in contact at the same temperature, and expanding or compressing the fluid in a way that would normally heat it or cool it; the heat will flow in or out instead.

For example: Stopper a *bicycle pump* and push the plunger in. It will get hot. Stick it in a bucket of water, and the heat will flow from the hot pump to the cold water. Now take it out and let the plunger up. The pump will get cold. Stick it in the water again, and heat will flow back into the pump. However, the amount of work you got back, when the plunger pushed its way up, was less than what you put in to push it down.

Now put the pump into the water first. Push the plunger down and you warm the water up a little, but the pump and water remain at about the same temperature. Let go: the pump and water will cool down a smidgen, and you'll get back a better fraction of the energy you put in. (In theory you could get back arbitrarily close to all of it, if the pump were frictionless and you moved it slowly enough to prevent buildup of temperature differences.)

Such an *isothermal* (without temperature drops) heat transfer is called *reversible*, because, as in our thought experiment, it works both ways without losing energy. The process of compression or expansion can also be reversible while warming up or cooling down, by virtue of insulating your fluid. You obviously don't want your heat leaking off into the environment. The engineer's jargon for a process that is absolutely insulated against heat flow is *adiabatic*. Neither kind of reversible process is possible with any technology like metal cylinders and pistons, but it showed Carnot the way to the theoretically optimal heat engine.

Here's how it works. We have two buckets, one with hot water and one with cold. We want to extract work by moving heat from the hot to the cold. We'll assume our bicycle pump is completely insulated when it is not in either bucket. Start with the plunger halfway out and the pump in the cold bucket.

1. (*Adiabatic compression*) Take the pump out of the cold bucket and push in the plunger until it is as hot as the hot bucket.
2. (*Isothermal expansion*) Put the pump into the hot bucket and let the plunger out halfway.
3. (*Adiabatic expansion*) Take the pump out and let the plunger the rest of the way out, its temperature falling to that of the cold bucket.
4. (*Isothermal compression*) Put the pump into the cold bucket and push the plunger in halfway.

Lather - Rinse - Repeat

You have put work into the pump in steps 1 and 4, but you get more out in steps 2 and 3. The reason: pressure rises and falls greatly during the adiabatic steps, but stays closer to constant during the isothermal ones. Thus step 2 gives you more work out than you put in step 4, while 1 and 3 more or less even out. Since you ended where you started, you can keep repeating as long the hot bucket is hotter than the cold bucket, getting work out every cycle.

Now, Carnot went on to note, every step is reversible. That means that you could do the whole thing backward, compressing in the hot bucket and expanding in the cold one. You'd put in more work than you got out, but you'd be pumping heat from the cold bucket to the hot one. In other words, a heat engine is a heat pump in reverse; and a reversible heat engine would pump the same amount of heat for a given amount of work, as it would produce work for the given amount of heat.

So, Carnot continued, imagine a heat engine more efficient than a reversible one. That is, one that produces more work for the same amount of heat. Well, just hook it to a reversible engine used as a pump, which can pump back from cold to hot all the heat you used from hot to cold, and you'll get some extra work out besides. In other words, get work out of nothing. And that, as Carnot stated with typical Gallic flair, is "repugnant to science and to common sense."

Thus the industrial revolution dawned. The revolutionary part was the beginning of a process, not the full course of one. Many of the trends set in course by the revolution continue today. One is the improvement of engines. The smaller and more powerful an engine, the more useful it is. The *Newcomen engine* could not be used in its original form for much more than pumping the water out of coal mines. It used too much coal to be economical very far from one. The Watt engines were useful as power sources for factories all over the country; and the Woolf engine was put on a wagon to replace a horse. By 1900, steam engines had become light enough to power a car; the Stanley Steamer gave gasoline engines a run for their money. But for mobile applications, heat engines using air and gaseous combustion products as a working fluid were the clear choice over steam, and have ruled the twentieth century. No useful airplane ever ran on steam. By the middle of the twentieth century, piston engines were being pushed out: of the most demanding applications by the gas turbine. Note that steam turbines had succeeded steam pistons in applications like ships through the first half of the century, and are still in use in stationary power plants today.

The reason is Carnot. In a stationary plant, your main concern is efficiency. If weight is no object, you can get just as high efficiency with steam as with any other working fluid. In fact, a steam power plant is notably more efficient than the gasoline engine in your car. However, in a car, light weight is the important thing. It is even more important in an airplane or a spaceship. The apotheosis of engines today is the *rocket engine*. Each ounce of a modern rocket engine produces more power than Newcomen's entire three machine.

Two overlapping sets of trends can be seen in the three-hundred-year history of engine development. First, within the development of any particular kind of engine, a rise of efficiency and power follows an exponentially increasing curve but then levels off as that kind of engine nears the theoretical limits imposed by thermodynamics. The second trend is that a constant rain of new engine types is being invented. For some figures of merit, such as power-to-weight ratio, or size of the smallest, or total power of the largest, and so on, the curves of the different types patch together to produce a single, unbroken, three-hundred-year trend line that is itself exponential.

For example, if we take steam pistons, steam turbines, gasoline pistons, gas turbines, and rocket engines, we can overlay them to get a smoothly increasing power to weight curve, stretching from Newcomen to the space shuttle. This is all somewhat qualitative since

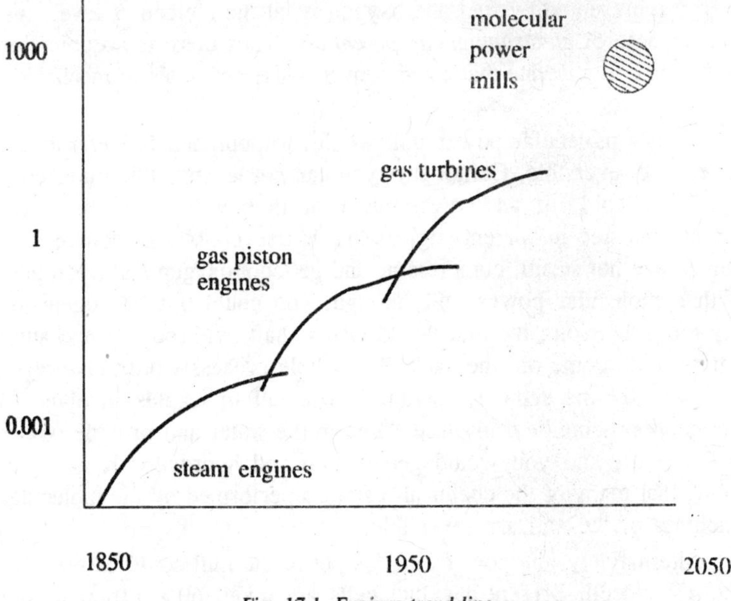

Fig. 17.1. Engines trend line.

one must take into account the cost of the engine, how much it pushed the technology, and what other options might have been taken in its stead. But the broad, general trend is unmistakable. The trend says that unless something really drastic happens to make the next thirty years totally different from the last three hundred, we'll have engines with the power to weight ratios, sizes, dimensional tolerances, and so forth that a predicts.

Another thing that has been increasing over the years is the efficiency of the engines. This does not mean that the laws of thermodynamics are being broken, but that a loophole is being used. The greater the difference between the hot and cold buckets, the more efficient the engine. Advances in materials, and some ingenious tricks like using a curtain of air to keep the flame in a jet engine from actually touching the walls of the combustion chamber, have allowed the temperature of the hot bucket to rise and rise.

With nanotechnology, however, it may be possible to finesse the matter altogether. No law, after all, says that no engine can be completely efficient; some electric motors get remarkably close. They do so by not being heat engines, but by using different principles. They use the *laws of electromagnetism* to translate voltage differences and current flows into mechanical forces. It turns out that the chemical energy represented by fuel and oxygen is, at the molecular level, just another case of *electromagnetic potentials*. Thus there is no objection in principle to a device that could convert those potentials to mechanical force.

Such a molecular power mill would, to approach full efficiency, have to be reversible. Carnot says you can not let something hot touch something cold, or an irreversible heat flow will occur. It is the combustion step in current engines that is irreversible, of course; you cannot take hot steam, compress it, and get cool oxygen and hydrogen. With a molecular power mill, though, you could put hydrogen and oxygen in the front, the mill would turn a shaft, and room-temperature water would come out the back. The whole process would take place at essentially the same temperature, with all molecules involved in the reaction being held in place. Feed in the water and provide power to the shaft, and you would get oxygen and hydrogen. It is worth noting that many of the chemical reactions performed by the molecular machines of the cell are reversible.

Alternatively, the power input/output of the mill could be electric, like a fuel cell; present-day fuel cells are about 60% efficient, but they are quite heavy. Power mills will represent the far end of the technology curve. Once the reversible mill is achieved, it could be combined with a hydrogen tank for an energy storage mechanism. Current-day car batteries store about one-sixth of a megajoule of energy per kilogram of weight; the best lithium batteries get up to half a megajoule. Gasoline provides 42 megajoules per kilogram, but you get only about 8 out of it with a car engine, the rest being discarded as heat. Hydrogen represents 120 megajoules.

In other words, if you are flying from New York to Miami in a light plane with a gasoline-burning piston engine, you'd need to carry 550 pounds of fuel, but a hydrogen powered reversible engine would need only 37 pounds of fuel.

There is only one problem: 37 pounds of hydrogen would fill a balloon about the size of Newcomen's original three-story steam engine. It would actually provide 450 pounds of lift! But it would he unmanageably awkward. The hydrogen needs to be kept in a more compact form; so, besides supplying the engine, nanotechnology will be needed for the fuel tank. Currently, hydrogen is stored compressed or liquified. The pressure tanks are heavy compared with the amount of hydrogen they can hold, and liquefaction is out of the comfort range of current technology, tripling the price of liquid hydrogen compared to the gas form. *Nanotech cooling* and *insulating technology* could help considerably. (Note that liquid hydrogen is quite light compared to other liquids, so the hydrogen tank would still be the same size as the gasoline tank; but it would weigh only 7% as much.

It's a lot harder to estimate the size of the engine, but we can get a ballpark figure by looking at estimates for molecular manufacturing systems. A fuel-burning engine is a manufacturing system at the molecular scale: it inputs the raw materials hydrogen and oxygen, and outputs water molecules. Using figures from Drexler's *Nanosystems*, we can estimate that a 100-kilowatt engine, as used by a car or light aircraft, would weigh in at about 50 grams (1.5 ounces).

Powering the Nano World

Newcomen's engine, occupying a full three-story building, supplied about as much power as the motor on a present-day lawn mower and consumed so much fuel it could only be used right next to a coal mine. The engines nanotechnology will build will make our current power plants look as bulky and inefficient—and as weak. It's one thing to realize that an engine matching the one in your car could be smaller and lighter than a pocketknife; it's another to realize that your pocketknife, if you wish, can have the power of fifty chainsaws. (And the reach of a chainsaw as well—a foldable, powered, woodcutting blade would be a fairly straightforward design with the materials and machines of molecular engineering.)

An amusing thing happened when he was doing the engineering calculations for the flying cars that appear later in the book. A collection of existing flying machine designs had made a few flying cars from fiction. Generally the serious fictional ones followed the same general form as the real aircraft. A few did not, seeming to follow the whimsy of the illustrator: the most recognizable of these is the Jetsons' car from the Hanna-Barbera cartoon. After doing the calculations, it was realized that only the *Jetsons* got it right; the engines would actually be the size of the silly little pods shown in the cartoons, if not smaller.

In virtually every application in which motors are used today, those motors will disappear. Fools and machines will still be powered, but the motors will become too small to see, part of the material. Objects will seem to be capable of moving by themselves, like animals. Your toothbrush won't have a motor in the handle, but many motors in each bristle. Instead of a single engine and wheels, a car might have hundreds of small legs, giving you a smooth, rock-steady ride over almost any surface. Engines will vanish, and will be everywhere.

18

DESTRUCTIVE LIMITATIONS

Thinking machines and *replicating assemblers* pose basic threats to people and to life on Earth. Today organisms have abilities far from the limits of the possible, and our machines are evolving faster than we are. Within a few decades they seem likely to surpass us. Unless we learn to live with them in safety, our future will likely be both exciting and short. We cannot hope to foresee all the problems ahead, yet by paying attention to the big, basic issues, we can perhaps foresee the greatest challenges and get some idea of how to deal with them. Entire books will no doubt be written on the coming social up heavals: What will happen to the global order when assemblers and automated engineering eliminate the need for most international trade? How will society change when individuals can live indefinitely? What will we do when replicating assemblers can make almost anything without human labour? What will we do when AI systems can think faster than humans? (And before they jump to the conclusion that people will despair of doing or creating anything, the authors may consider how runners regard cars, or how painters regard cameras.)

In fact, authors have already foreseen and discussed several of these issues. Each is a matter of uncommon importance, but more fundamental than any of them is the survival of life and liberty. After all, if life or liberty is obliterated, then our ideas about social problems will no longer matter.

THREAT FROM THE MACHINES

As described already, some of what replicating assemblers will do for us if we handle them properly. Powered by fuels or sunlight, they will be able to make almost anything (including more of them selves) from common materials. Living organisms are also powered by fuels or sunlight, and also make more of themselves from ordinary materials. But unlike assembler-based systems, they cannot make "*almost anything*."

Genetic evolution has limited life to a system based on DNA, RNA, and ribosomes, but *memetic evolution* will bring life-like machines based on *nanocomputers* and assemblers. It has already described, how assembler-built molecular machines will differ from the ribosome-built machinery of life. Assemblers will be able to build all that ribosomes can, and more; assembler-based replicators will there fore be able to do all that life can, and more. From an evolutionary point of view, this poses an obvious threat to otters, people, cacti, and ferns—to the rich fabric of the biosphere and all that we prize.

The early *transistorized computers* soon beat the most advanced vacuum tube computers because they were based on superior devices. For the same reason, early assembler-based replicators could beat the most advanced modern organisms. "*Plants*" with "*leaves*" no more efficient than today's solar cells could out-compete real plants, crowding the biosphere with an inedible foliage. Tough, omnivorous "bacteria" could out-compete real bacteria: they could spread like blowing pollen, replicate swiftly, and reduce the biosphere to dust in a matter of days. Dangerous replicators could easily be too tough, small, and rapidly spreading to stop—at least if we made no preparation. We have trouble enough controlling viruses and fruit flies.

Among the cognoscenti of nanotechnology, this threat has become known as the "*gray goo problem*." Though masses of uncontrolled replicators need not be gray or gooey, the term "*gray goo*" emphasizes that replicators able to obliterate life might be less inspiring than a single species of crabgrass. They might be "superior" in an evolutionary sense, but this need not make them valuable. We have evolved to love a world rich in living things, ideas, and diversity, so there is no reason to value gray goo merely because it could spread. Indeed, if we prevent it we will thereby prove our evolutionary superiority. The gray goo threat makes one thing perfectly clear: we cannot afford certain kinds of accidents with replicating assemblers.

Some of what advanced AI systems described already will do for us, if we handle them properly. Ultimately, they will embody the patterns of thought and make them flow at a pace no mammal's brain can match. AI systems that work together as people do will be able to out-think not just individuals, but whole societies. Again, the evolution of genes has left life stuck. Again, the evolution of memes by human beings—and eventually by machines—will advance our hardware far beyond the limits of life. And again, from an evolution any point of view this poses an obvious threat. Knowledge can bring power, and power can bring knowledge. Depending on their natures and their goals, advanced AI systems might accumulate enough knowledge and power to displace us, if we don't prepare properly,. And as with replicators, mere evolutionary "superiority" need not make the victors better than the vanquished by any standard but brute competitive ability. This threat makes one thing perfectly clear: we need to find ways to live with thinking machines, to make them law-abiding citizens.

Engines of Power

Certain kinds of replicators and *AI systems* may confront us with forms of hardware capable of swift, effective, independent action. But the novelty of this threat—coming from the machines themselves—must not blind us to a more traditional danger. Replicators and AI systems can also serve as great engines of power, if wielded freely by sovereign states.

Throughout history, states have developed technologies to extend their military power, and states will no doubt play a dominant role in developing replicators and AI systems. States could use replicating assemblers to build arsenals of advanced weapons, swiftly, easily, and in vast quantity. States could use special replicators directly to wage a sort of germ warfare—one made vastly more practical by programmable, computer-controlled "*germs*." Depending on their skills, AI systems could serve as weapon designers, strategists, or fighters. Military funds already support research in both molecular technology and artificial intelligence.

States could use assemblers or advanced AI systems to achieve sudden, destabilizing breakthroughs. As earlier discussed reasons for expecting that the advent of replicating assemblers will bring relatively sudden changes: Able to replicate swiftly, they could become abundant in a matter of days. Able to make almost anything, they could be programmed to duplicate existing weapons, but made from superior materials. Able to work with standard, well-understood components (atoms) they could suddenly build things designed in anticipation of the assembler breakthrough. These results of design-ahead could include programmable germs and other nasty novelties. For all these reasons, a state that makes the assembler breakthrough could rapidly create a decisive military force—if not literally over night, then at least with unprecedented speed. States could use *advanced AI* systems to similar ends. Automated engineering systems will facilitate design-ahead and speed assembler development. AI systems able to build better AI systems will allow an explosion of capability with effects hard to anticipate. Both AI systems and replicating assemblers will enable states to expand their military capabilities by orders of magnitude in a brief time.

Replicators can be more potent than nuclear weapons: to devastate Earth with bombs would require masses of exotic hardware and rare isotopes, but to destroy all life with replicators would require only a single speck made of ordinary elements. Replicators give nuclear war some company as a potential cause of extinction, giving a broader context to extinction as a moral concern. Despite their potential as engines of destruction, nanotechnology and AI systems will lend themselves to more subtle uses than do nuclear weapons. A bomb can only blast things, but nano machines and AI systems could be used to infiltrate, seize, change, and govern a territory or a world. Even the most ruthless police have no use for nuclear weapons, but they do have use for bugs, drugs, assassins, and other engines of power. With advanced technology, states will be able to consolidate their power over people.

Like *genes*, memes, organisms, and hardware, *states* have evolved. Their institutions have spread (with variations) through growth, fission imitation, and conquest. States at war fight like beasts, but using citizens as their bones, brains, and muscle. The coming breakthrough will confront states with new pressures and opportunities, encouraging sharp changes in how states behave. This naturally gives cause for concern. States have, historically, excelled at slaughter and oppression. In a sense, a state is simply the sum of the people making up its organ apparatus: their actions add up to make its actions. But the same might be said of a dog and its cells, though a dog is clearly more than just a clump of cells. Both dogs and states are evolved system with structures that affect how their parts behave. For thousands of years, dogs have evolved largely to please people, because they have survived and reproduced at human whim. For thousands of years, states have evolved under other selective pressures. Individuals have far more power over their dogs than they do over "their" states, Though states, too, can benefit from pleasing people, their very existence has depended on their capability for using people, whether as leaders, police, or soldiers. It may seem paradoxical to say that people have limited power over states: After all, aren't people behind a state's every action? But in democracies, heads of state bemoan their lack of power, representatives bow to interest groups, bureaucrats are bound by rules, and voters, allegedly in charge, curse the whole mess. The state acts and people affect it, yet no one can claim to control it. In totalitarian states, the apparatus of power has a tradition, structure, and inner logic that leaves no one free, neither the rulers nor the ruled. Even kings had to act in ways limited by the traditions of monarchy and the practicalities of power, if they were to remain kings. *States* are not human, though they are made of humans.

Despite this, history shows that change is possible, even change for the better. But changes always move from one semiautonomous, inhuman system to another—equally inhuman but perhaps more humane, In our hope for improvements, we must not confuse states that wear a human face with states that have humane institutions. Describing states as *quasi-organisms* captures only one aspect of a complex reality, yet it suggests how they may evolve in response to the coming breakthroughs. The growth of government power, most spectacular in totalitarian countries, suggests one direction. States could become more like organisms by dominating their parts more completely. Using replicating assemblers, states could fill the human environment with miniature surveillance devices. Using an abundance of speech-understanding *AI systems*, they could listen to everyone without employing half the population as listeners. Using nanotechnology like that proposed for cell repair machines, they could cheaply tranquilize, lobotomize, or otherwise modify entire populations. This would simply extend an all too familiar pattern. The world already holds governments that spy, torture, and drug; advanced technology will merely extend the possibilities.

But with advanced technology, states need not control people—they could instead simply *discard* people. Most people in most states, after all, function either as workers, larval workers, or worker-rearers, and most of these workers make, move, or grow things. A state with replicating assemblers would not need such work. What is more, advanced AI systems could replace engineers, scientists, administrators, and even leaders. The combination of nanotechnology and advanced AI will make possible intelligent, *effective robots*; with such robots, a state could prosper while discarding anyone, or even (in principle) everyone.

The implications of this possibility depend on whether the state exists to serve the people, or the people exist to serve the state.

In the first case, we have a state shaped by human beings to serve general human purposes; democracies tend to be at least rough approximations to this ideal. If a democratically controlled government loses its need for people, this will basically mean that it no longer needs to use people as bureaucrats or taxpayers. This will open new possibilities, some of which may prove desirable. In the second case, we have a state evolved to exploit human beings, perhaps along totalitarian lines. States have needed people as workers because human labour has been the necessary foundation of power. What is more, genocide has been expensive and troublesome to organize and execute. Yet, in this century totalitarian states have slaughtered their citizens by the millions. Advanced technology will make workers unnecessary and genocide easy. History suggests that *totalitarian states* may then eliminate people wholesale. There is some consolation in this. It seems likely that a state willing and able to enslave us biologically would instead simply kill us. The threat of advanced technology in the hands of governments makes one thing perfectly clear: we cannot afford to have an oppressive state take the lead in the coming breakthroughs.

The basic problems already outlined are obvious: in the future, as in the past, new technologies will lend themselves to accidents and abuse. Since replicators and thinking machines will bring great new powers, the potential for accidents and abuse will likewise be great. These possibilities pose genuine threats to our lives. Most people would like a chance to live and be free to choose how to live. This goal may not sound too utopian, at least in some parts of the world. It doesn't mean forcing everyone's life to fit some grand scheme; it chiefly means avoiding enslavement and death. Yet, like the achievement of a utopian dream, it will bring a future of wonders.

Given these life-and-death problems and this general goal, we can consider what measures might help us succeed. Our strategy must involve people, principles, and institutions, but it must also rest on tactics which inevitably will involve technology.

Trustworthy Systems

To use such powerful technologies in safety, we must make hard ware we can trust. To have trust, we must be able to judge technical facts accurately, an ability that will in turn depend partly on the quality of our institutions for judgment. More fundamentally, though, it will depend on whether trustworthy hardware is physically possible. This is a matter of the reliability of components and of systems.

Reliable components are often made even without assemblers to help. "*Reliable*" doesn't mean "*indestructible*"—anything will fail if placed close enough to a nuclear blast. It doesn't even mean "tough"—a television set may be reliable, yet not survive being bounced off a concrete floor. Rather, we call something reliable when we can count on it to work as designed. A reliable component need not be a perfect embodiment of a perfect design: it need only be a good enough embodiment of a cautious enough design. A bridge engineer may be uncertain about the strength of winds, the weight of traffic, and the strength of steel, but by assuming high winds, heavy traffic, and weak steel, the engineer can design a bridge that will stand. Unexpected failures in components commonly stem from material flaws. But assemblers will build components that have a negligible number of their atoms out of place—none, if need be. This will make them perfectly uniform and in a limited sense perfectly reliable. Radiation will still cause damage, though, because a cosmic ray can unexpectedly knock atoms loose from anything. In a small enough component (even in a modern computer memory device), a single particle of radiation can cause a failure.

But systems can work even when their parts fail; the key is redundancy. Imagine a bridge suspended from cables that fail randomly, each breaking about once a year at an unpredictable time. If the bridge falls when a cable breaks, it will be too dangerous to use. Imagine, though, that a broken cable takes a day to fix (because skilled repair crews with spare cables are on call), and that, though it takes five cables to support the bridge, there are actually *six*. Now if one cable breaks, the bridge will still stand. By clearing traffic and then replacing the failed cable, the bridge operators can restore safety. To destroy this bridge, a second cable must break in the same day as the first. Supported by six cables, each having a one-in-365 daily chance of breaking, the bridge will likely last about ten years. While an improvement, this remains terrible. Yet a bridge with ten cables (five needed, five extra) will fall only if six cables break on the same day: the suspension system is likely to last over ten million years. With fifteen cables, the expected lifetime is over ten thousand times the age of the Earth. *Redundancy* can bring an exponential explosion of safety.

Redundancy works best when the redundant components are truly independent. If we don't trust the design process, then we must use components designed independently; if a bomb, bullet, or cosmic ray may damage several neighbouring parts, then we must spread redundant parts more widely.

Engineers who want to supply reliable transportation between two islands shouldn't just add more cables to a bridge. They should build two well-separated bridges using different designs, then add a tunnel, a ferry, and a pair of inland airports. Computer engineers also use redundancy. Stratus Computer Inc., for example, makes a machine that uses four central processing units (in two pairs) to do the work of one, but to do it vastly more reliably. Each pair is continually checked for internal consistency, and a failed pair can be replaced while its twin carries on. An even more powerful form of redundancy is *design diversity*. In computer hardware, this means using several computers with *different designs*, all working in parallel. Now *redundancy* can correct not just for failures in a piece of hardware, but for errors in its design.

Much has been made of the problem of writing large, error-free programs; many people consider such programs impossible to develop and debug. But researchers at the UCLA Computer Science Department have shown that design diversity can also be used in software: several programmers can tackle the same problem independently, then all their programs can be run in parallel and made to vote on the answer. This multiplies the cost of writing and running the program, but it makes the resulting software system resistant to the bugs that appear in some of its parts. We can use redundancy to control replicators. Just as repair machines that compare multiple DNA strands will be able to correct mutations in a cell's genes, so replicators that compare multiple copies of their instructions (or that use other effective error-correcting systems) will be able to resist mutation in these "*genes*." Redundancy can again bring an exponential explosion of safety.

We can build systems that are extremely reliable, but this will entail costs. Redundancy makes systems heavier, bulkier, more ex pensive, and less efficient. *Nanotechnology*, though, will make most things far lighter, smaller, cheaper, and more efficient to begin with. This will make redundancy and reliability more practical. Today, we are seldom willing to pay for the safest possible systems; we tolerate failures more-or-less willingly and seldom consider the real limits of reliability. This biases judgments of what can be achieved. A psychological factor also distorts our sense of how reliable things can be made: failures stick in our minds, but everyday successes draw little attention. The media amplify this tendency by reporting the most dramatic failures from around the world, while ignoring the endless and boring successes. Worse yet, the components of redundant systems may fail in visible ways, stirring alarums: imagine how the media would report a snapped bridge cable, even if the bridge were the super-safe fifteen-cable model described above. And since each added redundant component adds to the chance of a component failure, a system's reliability can seem worse even as it approaches perfection.

Appearances aside, redundant systems made of abundant, flawless components can often be made almost perfectly reliable. Redundant systems spread over wide enough spaces will survive even bullets and bombs. But what about design errors? Having a dozen redundant parts will do no good if they share a fatal error in design. Design diversity is one answer, good testing is another. We can reliably evolve good designs without being reliably good designers: we need only be good at testing, good at tinkering, and good at being patient. Nature has evolved working molecular machinery through entirely mindless tinkering and testing. Having minds, we can do as well or better. We will find it easy to design reliable hardware if we can develop reliable automated engineering systems. But this raises the wider issue of developing trustworthy artificial intelligence systems. We will have little trouble making AI systems with reliable hardware, but what about their software?

Like present AI systems and human minds, advanced *AI systems* will be synergistic combinations of many simpler parts. Each part will be more specialized and less intelligent than the system as a whole. Some parts will look for patterns in pictures, sounds, and other data and suggest what they might mean. Other parts will compare and judge the suggestions of these parts. Just as the pattern recognizers in the human visual system suffer from errors and optical illusions, so will the pattern recognizers in AI systems. (Indeed, some advanced machine vision systems already suffer from familiar optical illusions.) And just as other parts of the human mind can often identify and compensate for illusions, so will other parts of AI systems.

As in human minds, *intelligence* will involve mental parts that make shaky guesses and other parts that discard most of the bad guesses before they draw much attention or affect important decisions. Mental parts that reject action ideas on ethical grounds correspond to what we call a conscience. AI systems with many parts will have room for redundancy and design diversity, making reliability possible.

A genuine, flexible AI system must evolve ideas. To do this, it must find or form hypotheses, generate variations, test them, and then modify or discard those found inadequate. Eliminating any of these abilities would make it stupid, stubborn, or insane ("*Durn machine can't think and won't learn from its mistakes—junk it!*"). To avoid becoming trapped by initial misconceptions, it will have to consider conflicting views, seeing how well each explains the data, and seeing whether one view can explain another.

Scientific communities go through a similar process. And in a paper called "The Scientific Community Metaphor," William A. Kornfeld and Carl Hewitt of the MIT Artificial Intelligence Laboratory suggest that AI researchers model their programs still more closely on the evolved structure of the scientific community. They point to the pluralism of science, to its diversity of competing proposers, supporters, and critics. Without proposers, ideas cannot appear; without supporters, they cannot grow; and without critics to weed them, bad ideas can crowd out the good. This holds true in science, in technology, in AI systems, and among the parts of our own minds.

Having a world full of diverse and *redundant proposers*, supporters, and critics is what makes the advance of science and technology reliable. Having more proposers makes good proposals more common; having more critics makes bad proposals more vulnerable. Better, more numerous ideas are the result. A similar form of redundancy can help AI systems to develop sound ideas. People sometimes guide their actions by standards of truth and ethics, and we should be able to evolve AI systems that do likewise, but more reliably. Able to think a million times faster than us, they will have more time for second thoughts. It seems that AI systems can be made trustworthy, at least by human standards.

AI systems to individual human minds are often compared but the resemblance need not be close. A system that can mimic a person may need to be person like, but an automated engineering system probably doesn't. One proposal (called an *Agora system*, after the Greek term for a meeting and market place) consist of many independent pieces of software that interact by offering one another services in exchange for money. Most pieces would be simple minded specialists, some able to suggest a design change, and others able to analyze one.

Much as Earth's ecology has evolved extraordinary organisms, so this computer economy could evolve extraordinary designs—and perhaps in a comparably mindless fashion. What is more, since the system would be spread over many machines and have parts written by many people, it could be diverse, robust, and hard for any group to seize and abuse. Eventually, one way or another, automated engineering systems will be able to design things more reliably than any group of human engineers can today. Our challenge will be to design them correctly. We will need human institutions that reliably develop reliable systems. Human institutions are evolved artificial systems, and they can often solve problems that their individual members cannot. This makes them a sort of "*artificial intelligence system.*" Corporations, armies, and research laboratories all are examples, as are the looser structures of the market and the scientific community.

Even governments may be seen as artificial intelligence systems—gross, sluggish, and befuddled, yet superhuman in their sheer capability. And what are constitutional checks and balances but an attempt to increase a government's reliability through institutional diversity and redundancy? When we build intelligent machines, we will use them to check and balance one another. By applying the same principles, we may be able to develop reliable, technically oriented institutions having strong checks and balances, then use these to guide the development of the systems we will need to handle the coming breakthroughs.

Tactics for the Assembler Breakthrough

Some force in the world (whether trustworthy or not) will take the lead in developing assemblers; call it the "*leading force.*" Because of the strategic importance of assemblers, the leading force will presumably be some organization or institution that is effectively controlled by some government or group of governments. To simplify matters, pretend for the moment that we (the good guys, attempting to be wise) can make policy for the leading force. For citizens of democratic states, this seems a good attitude to take. What should we do to improve our chances of reaching a future worth living in? What can we do?

We can begin with what must not happen: we must not let a single replicating assembler of the wrong kind be loosed on an unprepared world. Effective preparations seem possible, but it seems that they must be based on assembler-built systems that can be built only after dangerous replicators have already become possible. Design-ahead can help the leading force prepare, yet even vigorous, foresighted action seems inadequate to prevent a time of danger. The reason is straightforward: dangerous replicators will be far simpler to design than systems that can thwart them, just as a bacterium is far simpler than an immune system. We will need tactics for containing nanotechnology while we learn how to tame it.

One obvious tactic is *isolation*: the leading force will be able to contain replicator systems behind multiple walls or in laboratories in space. Simple replicators will have no intelligence, and they won't be designed to escape and run wild. Containing them seems no great challenge. Better yet, we will be able to design replicators that cannot escape and run wild. We can build them with counters (like those in cells) that limit them to a fixed number of replications. We can build them to have requirements for special synthetic "*vitamins*," or for bizarre environments found only in the laboratory. Though replicators could be made tougher and more voracious than any modern pests, we can also make them useful but harmless. Because we will design them from scratch, replicators need not have the elementary survival skills that evolution has built into living cells. Further, they need not be able to evolve. We can give replicators redundant copies of their "genetic" instructions, along with repair mechanisms to correct any mutations. We can design them to stop working long before enough damage accumulates to make a lasting mutation a significant possibility. Finally, we can design them in ways that would hamper evolution even if mutations could occur.

Experiments show that most computer programs seldom respond to mutations by changing slightly; instead, they simply fail. Because they cannot vary in useful ways, they cannot evolve. Unless they are specially designed, replicators directed by nano computers will share this handicap. Modern organisms are fairly good at evolving partly because they descend from ancestors that evolved. They are evolved to evolve; this is one reason for the complexities of sexual reproduction and the shuffling of chromosome segments during the production of sperm and egg cells. We can simply neglect to give replicators similar talents. It will be easy for the leading force to make replicating assemblers useful, harmless, and stable. Keeping assemblers from being stolen and abused is a different and greater problem, because it will be a game played against intelligent opponents. As one tactic, we can reduce the incentive to steal assemblers by making them available in safe forms. This will also reduce the incentive for other groups to develop assemblers independently. The leading force, after all, will be followed by trailing forces.

Limited Assemblers

It is pointed out that we will be able to make *assembler systems* that act like seeds, absorbing sunlight and ordinary materials and growing to become almost anything. These special-purpose systems will not replicate themselves, or will do so only a fixed number of times. They will make only *what* they were programmed to make, *when* they are told to make it. Anyone lacking special assembler-built tools would be unable to reprogram them to serve other purposes. Using limited assemblers of this sort, people will be able to make as much as they want of whatever they want, subject to limits built into the machines. If none is programmed to make nuclear weapons, none will; if none is programmed to make dangerous replicators, none will. If some are programmed to make houses, cars, computers, toothbrushes, and whatnot, then these products can become cheap and abundant.

Machines built by limited assemblers will enable us to open space, heal the biosphere, and repair human cells. Limited assemblers can bring almost unlimited wealth to the people of the world. This tactic will ease the moral pressure to make unlimited assemblers available immediately. But limited assemblers will still leave legitimate needs unfulfilled. Scientists will need freely programmable assemblers to conduct studies; engineers will need them to test designs. These needs can be served by sealed assembler laboratories.

Sealed Assembler Laboratories

Picture a computer accessory the size of your thumb, with a state-of-the-art plug on its bottom. Its surface looks like boring gray plastic, imprinted with a serial number, yet this sealed assembler lab is an assembler-built object that contains many things. Inside, sitting above the plug, is a large *nanoelectronic computer* running advanced molecular-simulation software (based on the software developed during assembler development). With the assembler lab plugged in and turned on, your assembler-built home computer displays a three-dimensional picture of whatever the lab computer is simulating, rep resenting atoms as coloured spheres. With a joystick, you can direct the simulated assembler arm to build things. Programs can move the arm faster, building elaborate structures on the screen in the blink of an eye. The simulation always works perfectly, because the nanocomputer cheats: as you make the simulated arm move simulated molecules, the computer directs an actual arm to move actual molecules. It then checks the results whenever needed to correct its calculations. The end of this thumb-sized object holds a sphere built in many concentric layers.

Fine wires carry power and signals through the layers; these let the nano computer in the base communicate with the devices at the sphere's center. The outermost layer consists of sensors. Any attempt to remove or puncture it triggers a signal to a layer near the core.

The next layer in is a thick spherical shell of pre stressed diamond composite, with its outer layers stretched and its inner layers compressed. This surrounds a layer of thermal insulator which in turn surrounds a peppercorn-sized spherical shell made up of microscopic, carefully arranged blocks of metal and oxidizer. These are laced with electrical igniters. The outer sensor layer, if punctured, triggers these igniters. The metal-and-oxidizer demolition charge then burns in a fraction of a second, producing a gas of metal oxides denser than water and almost as hot as the surface of the Sun. But the blaze is tiny; it swiftly cools, and the diamond sphere confines its great pressure. This demolition charge surrounds a smaller composite shell, which surrounds another layer of sensors, which can also trigger the demolition charge. These sensors surround the cavity which contains the actual sealed assembler lab.

These elaborate precautions justify the term "*sealed.*" Someone outside cannot open the lab space without destroying the contents, and *no assemblers* or *assembler-built structures can escape from within.* The system is designed to let out information, but not dangerous replicators or dangerous tools. Each sensor layer consists of many redundant layers of sensors, each *intended* to detect any possible penetration, and each making up for possible flaws in the others. Penetration, by triggering the demolition charge, raises the lab to a temperature beyond the melting point of all possible substances and makes the survival of a dangerous device impossible. These *protective mechanisms* all gang up on something about a millionth their size—that is, on whatever will fit in the lab, which provides a spherical work space no wider than a human hair. Though small by ordinary standards, this work space holds room enough for millions of assemblers and thousands of trillions of atoms. These sealed labs will let people build and test devices, even voracious replicators, in complete safety.

Children will use the atoms inside them as a construction set with almost unlimited parts. *Hobbyists* will exchange programs for building various gadgets. Engineers will build and test new nanotechnologies. Chemists, materials scientists, and biologists will build apparatus and run experiments. In labs built around biological samples, biomedical engineers will develop and test early cell repair machines. In the course of this work, people will naturally develop useful designs, whether for computer circuits, strong materials, medical de vices, or whatever. After a public review of their safety, these things could be made available outside the sealed labs by programming limited assemblers to make them. Sealed labs and limited assemblers will form a complementary pair: The first will let us invent freely; the second will let us enjoy the fruits of our invention safely. The chance to pause between design and release will help us avoid deadly surprises. Sealed assembler labs will enable the whole of society to apply its creativity to the problems of nanotechnology. And this will speed our preparations for the time when an independent force learns how to build something nasty.

Hiding Information

In another tactic for buying time, the leading force can *attempt* to burn the bridge it built from bulk to *molecular technology*. This means destroying the records of how the first assemblers were made (or making the records thoroughly inaccessible). The leading force may be able to develop the first, crude assemblers in such a way that no one knows the details of more than a small fraction of the whole system. Imagine that we develop assemblers by the route. The protein machines that we use to build the first crude assemblers will then promptly become obsolete. If we destroy the records of the protein designs, this will hamper efforts to duplicate them, yet will not hamper further progress in nanotechnology.

If sealed labs and limited assemblers are widely available, people will have little scientific or economic motivation to redevelop nanotechnology independently, and burning the bridge from bulk technology will make independent development more difficult. Yet these can be no more than delaying tactics. They won't stop independent development; the human urge for power will spur efforts which will eventually succeed. Only detailed, universal policing on a totalitarian scale could stop independent development indefinitely, lithe policing were conducted by anything like a modern government, this would be a cure roughly as dangerous as the disease. And even then, would people maintain perfect vigilance *forever*? It seems that we must eventually learn to live in a world with untrustworthy replicators. One sort of tactic would be to hide behind a wall or to run far away. But these are brittle methods: dangerous replicators might breach the wall or cross the distance, and bring utter disaster. And, though walls can be made proof against small replicators, no fixed wall will be proof against large-scale, organized malice. We will need a more robust, flexible approach.

Active Shield

It seems that we can build *nanomachines* that act somewhat like the white blood cells of the human immune system: devices that can fight not just bacteria and viruses, but dangerous replicators of all sorts. Call an *automated defense* of this sort an *active shield*, to distinguish it from a fixed wall. Unlike ordinary engineering systems, reliable active shields must do more than just cope with nature or clumsy users. They must also cope with a far greater challenge—with the entire range of threats that intelligent forces can design and build under prevailing circumstances. Building and improving prototype shields will be akin to running both sides of an arms race on a laboratory scale. But the goal here will be to seek the minimum requirements for a defense that reliably prevails. Dr. Lenat and his Eurisko program evolved successful fleets to fight according to the rules of a naval-warfare simulation game. In a similar way, we can make into a game the deadly serious effort to develop reliable shields, using sealed assembler labs of various sizes as playing fields.

We can turn loose a horde of engineers, computer hackers, biologists, hobbyists, and automated engineering systems, all invited to pit their systems against one another in games limited only by the initial conditions, the laws of nature, and the walls of the sealed labs. These competitors will evolve threats and shields in an open-ended series of micro battles. When replicating assemblers have brought abundance, people will have time enough for so important a game. Eventually we can test promising shield systems in Earthlike environments in space. Success will make possible a system able to protect human life and Earth's biosphere from the worst that a fistful of loose replicators can do.

Is Success Possible?

With our present uncertainties, we cannot yet describe either threats or shields with any accuracy. Does this mean we can't have any confidence that effective shields are possible? Apparently we can; there is a difference, after all, between knowing that something is possible and knowing how to do it. And in this case, the world holds examples of *analogous successes*. There is nothing fundamentally novel about defending against invading replicators; life has been doing it for ages. *Replicating assemblers*, though unusually potent, will be physical systems not unlike those we already know. Experience suggests that they can be controlled.

Viruses are molecular machines that invade cells; cells use molecular machines (such as *restriction enzymes* and *antibodies*) to defend against them. Bacteria are cells that invade organisms; organisms use cells (such as white blood cells) to defend against them. Similarly, societies use police to defend against criminals and armies to defend against invaders. On a less physical level, minds use meme systems such as the scientific method to defend against nonsense, and societies use institutions such as courts to defend against the power of other institutions. The biological examples in the last paragraph show that even after a billion-year arms race, molecular machines have maintained successful defenses against molecular replicators. Failures have been common too, but the successes do indicate that defense is possible. These successes suggest that we can indeed use nanomachines to defend against nanomachines. Though assemblers will bring many advances, there seems no reason why they should permanently tip the balance against defense.

The examples given above—some involving viruses, some involving institutions—are diverse enough to suggest that successful defense rests on general principles. One might ask, Why do all these defenses succeed? But turn the question around: Why should they fail? Each *conflict pits* similar systems against each other, giving the attacker no obvious advantage. In each conflict, moreover, the attacker faces a defense *that is well established.* The defenders fight on home ground, giving them advantages such as prepared positions, detailed local knowledge, stockpiled resources, and *abundant allies*—when the immune system recognizes a germ, it can mobilize the resources of an entire body. All these advantages are general and basic, having little to do with the details of a technology. We can give our active shields the same advantages over dangerous replicators. And they need not sit idle while dangerous weapons are amassed, any more than the immune system sits idle while bacteria multiply.

It would be hard to predict the outcome of an open-ended arms race between powers equipped with replicating assemblers. But be fore this situation can arise, the leading force seems likely to acquire a temporary but overwhelming military advantage. If the outcome of an arms race is in doubt, then the leading force will likely use its strength to ensure that no opponents are allowed to catch up. If it does so, then active shields will not have to withstand attacks backed by the resources of half a continent or half a solar system; they will instead be like a police force or an immune system, facing attacks backed only by whatever resources can be gathered in secret within the protected territory.

In each case of successful defense, the attackers and the shields have developed through broadly similar processes. The immune system, shaped by genetic evolution, meets threats also shaped by *genetic evolution*. Armies, shaped by human minds, also meet similar threats. Likewise, both active shields and dangerous replicators will be shaped by memetic evolution. But if the leading force can develop automated engineering systems that work a million fold faster than human engineers, and if it can use them for a single year, then it can build active shields based on a million years worth of engineering advance. With such systems we may be able to explore the limits of the possible well enough to build a reliable shield against all physically possible threats.

Even without our knowing the details of the threats and the shields, there seems reason to believe that shields are possible. And the examples of memes controlling memes and of institutions con trolling institutions also suggest that *AI systems* can control *AI systems*. In building active shields, we will be able to use the power of replicators and AI systems to multiply the traditional advantages of the defending force: we can give it overwhelming strength through abundant, replicator-built hardware with designs based on the equivalent of a million-year lead in technology. We can build active shields having strength and reliability that will put past systems to shame. Nanotechnology and artificial intelligence could bring the ultimate tools of destruction, but they are not inherently destructive. With care, we can use them to build the ultimate tools of peace.

19

NANOBIOTECHNOLOGY

In this chapter practical uses for *current* biotechnology, will be dealt and about what tools biotechnology can present toward making structures for molecular nanotechnology. Protein structures will be primarily dealt, although there are many other structural components of cells that might be useful for nanotechnology. In particular, the physical-plant of a cell in terms of the design of the structural motifs of proteins, which could provide useful clues for designing nano sized structures are to be discussed including the tools (found in cells) that are used for receiving molecules from other sources, and holding molecules in storage until they are needed. How manufacturing control in a cell is enabled, and how inventory control is done in the cell will also be discussed touching about the management information systems in a cell, and manufacturing methods in a cell.

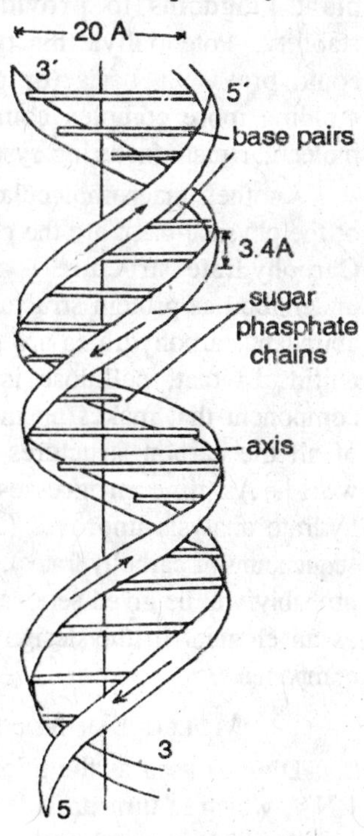

Fig. 19.1. Structure of DNA.

It is well known that DNA controls the metabolic processes of the cells by means of transcription to form RNA molecules, which are then translated to make protein molecules that embody the information previously stored in the

DNA. The protein molecules, among other functions performed in the cell, work through a feedback loop to control the expression a replication of the DNA.

Structural Proteins

The physical-plant of the cell, and the components that surround the cell (termed the *extracellular matrix*), consist of structural proteins. Examples are the proteins collagen and keratin. *Collagen*, which is found in skin, adopts a triple helical structure. *Keratin* also has a triple helical structure. Collagen is, in fact, a repeating polymer of just three amino acids, with a few special amino acids inserted to provide for cross-linking and stability. This type of cross-linking is found throughout the animal and plant kingdoms to provide structural stability. Potentially, this type of motif could provide a basis (or platform) for building more complex components in a molecular manufacturing system.

Fig. 19.2. Collagen is a structural protein found in cells.

Another macromolecular component of the physical-plant are the *carbohydrates*. Carbohydrate structure is not as well-understood as protein structure. The most abundant carbohydrates are cellulose and chitin. In fact, cellulose is a structural component that makes up more than half of all the carbon structures found in the world. As time progresses and carbohydrate analysis improves (especially the sequencing of carbohydrates), carbohydrate probably will be given serious consideration as an element in the design of structural components.

Molecular Manufacturing within the Cell

DNA is used as the information source for transcribing messenger RNA, which in turn provides the information to produce proteins. One way to produce novel proteins is to modify the DNA, and thus, modify the messenger RNA. Novel DNAs can be created by chemical synthesis of DNA, or by selection and isolation of natural mutations.

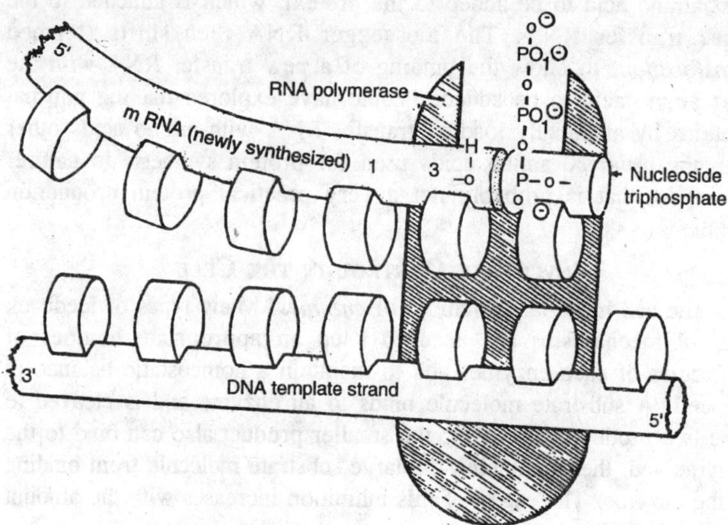

Fig. 19.3. Role of RNA polymerase in translation.

Once the message has been created for the manufacture of a protein, a protein of nearly any imaginable amino acid sequence can be manufactured. The *messenger RNA* binds to a complementary section of the two *transfer RNA* molecules bound to the ribosome. One transfer RNA contains the growing protein chain, which is then joined to the

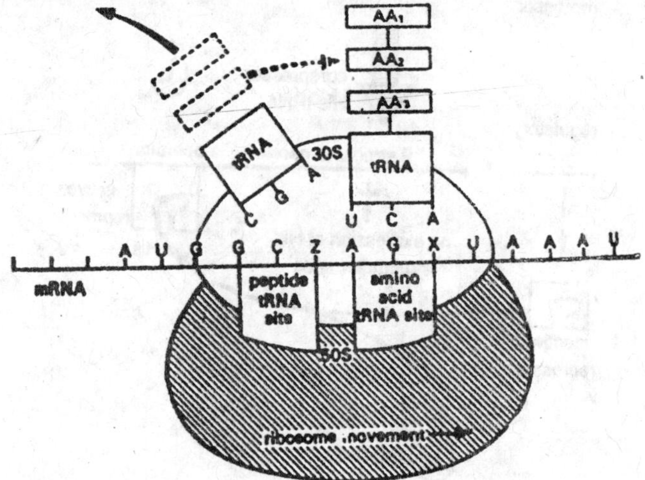

Fig. 19.4. Formation of polypeptide chain.

next amino acid to be added to the protein, which is attached to the other transfer RNA. The messenger RNA then shifts (termed *translocation*) to allow the binding of a new transfer RNA with the next amino acid to be added. People have explored making unusual proteins by artificially loading transfer RNA with amino acids other than the usual 20 amino acids used for protein synthesis in nature. However, that is probably not a very practical protein production technique.

Inventory Control in the Cell

The cell has a large number of *enzymes*. Many types of feedback control mechanisms are used to keep an appropriate number of molecules of each enzyme, and to maintain a homeostatic balance in the cell. A substrate molecule binds to an enzyme and is cleaved to give two products. However, the smaller product also can bind to the enzyme and, thereby, inhibit the large substrate molecule from binding to the enzyme. The extent of this inhibition increases with the amount

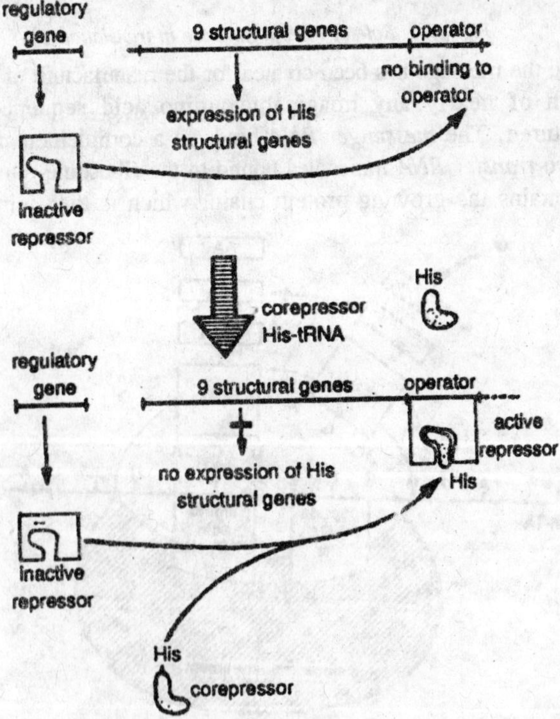

Fig. 19.5. Corepressor-repressor complex.

of the product that has already been produced. Thus, product binding to the enzyme effectively limits the concentration of the product that can be produced. This is a very common feedback mechanism in biosynthesis. The presence of many thousands of enzymes, each having a variety of feedback loops, leads to a very complicated process of regulation. Considering these feedback loops leads to questions such as how to design a protein appropriately so that the binding of some reaction product gives the desired level of feedback.

Recent Technology for Producing Proteins

Proteins and peptides can be obtained in a variety of ways. One of these is *classical solution-phase chemical synthesis*. In this case, amino acids are modified with appropriate protecting groups to protect reactive portions of the amino acid, which if allowed to react, would create unwanted products, and then dissolved in solution. A reagent is added that causes two amino acids to combine, forming a peptide bond between the amino acids, and to exclude a water molecule. This is the classical approach for *small peptides*.

To make *larger peptides* (up to approximately 40 amino acid residues long), the only practical synthetic approach is *solid-phase synthesis*. Here, the first amino acid is attached (*covalently bonded*) to a solid support, and one amino acid (again with appropriate chemically reactive groups protected) is added at a time until the desired chain has been built, still attached to the solid support. The synthetic product is then cleaved from the solid support, stripped of the protecting groups, and extensively purified to give the final product.

Fragment condensation is a variation of these two techniques. Fragments of the desired peptide are made on a solid support, and then later combined in solution. This has some of the advantages of both methods, although it also has disadvantages (principally that the fragments often are not very soluble so that they cannot be coupled in solution).

In addition to these chemical approaches, several biological approaches can also be used. Proteins of interest can be purified from natural sources. Or, the gene for a protein of interest can be inserted into cells, the cells can be grown in culture to express large quantities of that protein, and the protein can be purified from the cultured cells. Some proteins can be purified from fermentation broths of cells.

Semisynthesis is a combination of the previously discussed methods. For example, a protein of interest could be obtained by biological methods, and the split into fragments by using chemical or biochemical methods. Changes could be made to one or more fragments by using the methods of organic chemistry, and then the fragments could be recombined to form a partially synthetic molecule. The advantages of using the synthetic approaches include the ability to introduce (into the molecule) segments that are not normally found in nature. This gives the molecule conformational (or physical) properties that could not be achieved with a molecule taken directly from a cell.

In general, to make *small peptides* (on the order of two to eight amino acid residues—often the size that is used in the pharmaceutical industry) the most economical approach is to use solution methods. To make peptides of 20 to 30 residues, solid-phase is almost the only way to do rapid development work. Beyond this size, it is generally necessary to use *biological synthesis* or *semisynthesis*.

Protein Structures Begin with Linear Chains of Amino Acids

Protein molecules are sometimes depicted as *amorphous blobs*. Peptides and proteins are chains of amino acids. One of the main characteristics of amino acids is that they have *handedness*. The chemical term for that is *chirality*. The side chain (referred to as an *R group*) is shown extending to the left of the central (*alpha*) carbon atom. The 20 amino acids used in biological protein synthesis have 20 different R groups, which give the amino acids their different properties. The

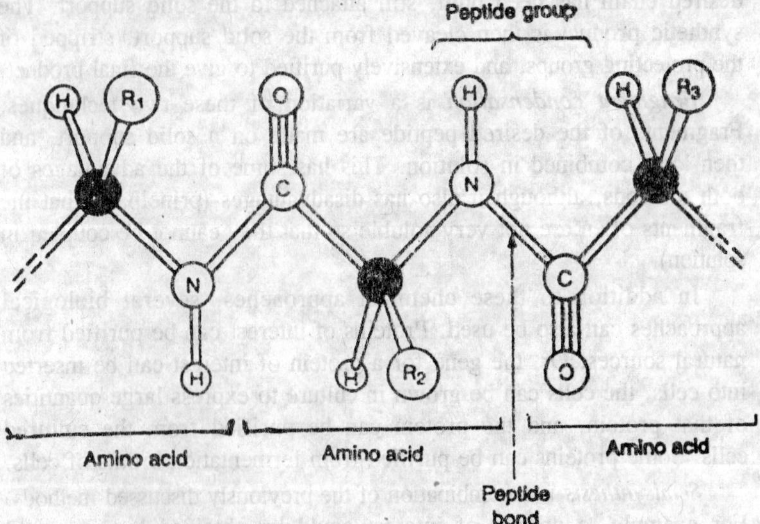

Fig. 19.6. The arrangement of atoms in amino acids joined by peptide bonds.

amino group containing a nitrogen atom is on the far side. Linking these amino acids together to form a chain, excluding a molecule of water for each peptide bond formed, gives a polypeptide. Conventionally, peptides of 50 or more amino acid residues (about the size of *insulin*) are called *proteins*.

Complex Proteins

In addition to the *peptide bonds* linking amino acids into a linear array, other structural elements are important features of protein structures. These elements contribute to folding the linear chain in three dimensions. One of these is *hydrogen bonding*. Here, a relatively weak bond is formed between certain hydrogen atoms and an oxygen that is distant along the linear sequence, but near in three-dimensions as the chain folds. A large number of these weak bonds can lend significant stability to a protein structure.

A more stabilizing element that is not found in all proteins is a *covalent disulfide bridge* providing a cross-link between two sections of the molecule that are again distant along the linear sequence, but spatially near as the chain folds.

There are 20 amino acids which are used in proteins, and they have very different properties. One group is *hydrophobic* (that is, they do not like water and can be described as "oily"). Another set is very bulky, and also hydrophobic. Another set is electrically charged and *hydrophilic* (that is, they like water and they make very good hydrogen bonds). Another set is *polar* (which means that they like water, but they are not charged). A few amino acids also make unique conformational contributions to the structure of the protein molecule. For example, proline is a cyclic amino acid often found in turns (where the protein chain moves out from the interior of the globular structure, makes a turn, and moves back inside).

Secondary Structure in Proteins

As polypeptides become larger, the protein structure becomes more complete and distinct *secondary* and *tertiary* structures are found. One common secondary structure feature is for the chain to form a *helix*. Helices of different types can be described by the pitch of the helix, which is related to the number of amino acid residues per turn of the helix, and to the distance along the helix axis covered by each residue.

As illustrated several helices with different pitches. The example of the left is a case where the helical pitch is *zero* and the peptide is a covalent closed circle (in this case, of five residues). The next helix has four residues turn, and is similar to the *alpha helix* found in many proteins. The alpha helix a stable, rod like structure in which the CO of one amino acid residue is hydrogen bonded to the NH four residues ahead in the linear sequence. This forms tight coil with 3.6 residues per turn.

If the helix is stretched further, it will form 3_{10} helix, with three residues per turn. If the helix is less stretched, it will form a pi helix, with 4.4 residues per turn. These helices are less favourable energetically than the alpha helix, and are seldom seen in protein structures. The potential to form hydrogen bonds is the major reason that polypeptides form helices. The strength of the hydrogen bonds made possible by the different helix geometries determines why some potential helix structures are more stable than others.

If the helix is stretched further so that there are only two residues per turn, the result is not a helix, but an extended structure termed the *beta strand*, which is as extended as a peptide chain can get.

If two beta strands are placed together, a *beta sheet* is formed. One of the interesting properties of the beta sheet is that it has two sides. The side chains of consecutive amino acids alternate in the direction to which they are presented. Thus, the side chains of alternate amino acid residues point either down or up. This allows the presentation of different faces of this structure to different environments within the protein. For example, if the upper side of the beta sheet were exposed to water, the side chains on the upper side of the beta sheet would probably be largely *hydrophilic* (charged or polar) while those on the bottom (pointing to the interior of the protein) would likely be *hydrophobic*. Having the hydrophilic side of the beta sheet in contact with water and the hydrophobic side in contact with a hydrophobic environment stabilizes the bet structure.

Many *hydrogen bonds* are formed between adjacent strands of a beta sheet and these help to stabilize the structure. In general, beta structures are found the interior of proteins. They tend to have more hydrophobic residues than do alpha helices, although there are some alpha helices that are fully buried wit the interiors of proteins.

An example of a beta structure is *silk*. Silk fibers are rich in alanine glycine residues. These fibers are made of *beta strands* in which alanine residence are on one side of the beta sheet and glycine residues are on the other. The beta sheets are packed on top of one another such that alanine residues on sheet face alanine residues on the next sheet, and glycine residues face glycine residues. The opposing alanine residues keep their respective beta sheets from moving with respect to each other, because moving the sheets across one another would cause the alanine side chains to collide. That is why silk does not stretch, although it is flexible perpendicular to the beta sheets.

One structure that allows the protein chain to abruptly change direction is called the *beta turn* structure. Here, the protein chain enters the turn from the top right, swings around, and continues to the bottom right. These structures are seen very frequently (in about 20 to 25 percent of the residues) in proteins and allow the chain to change direction 180 degrees.

Structural Motifs in Proteins

These elements of secondary structure (*alpha helices*, *beta strands*, and *beta turns*) combine to form a number of *structural motifs*. These same structural motifs are seen in very diverse proteins. The proteins may often have very different functions, yet they are very similar in terms of the organization of the elements of secondary structure into structural motifs. The most prominent feature consists of the four alpha helices, which are oriented in a parallel fashion with adjacent pairs of helices antiparallel (the helices point in opposite directions) to each other. The interior of the four-helix bundle is hydrophobic and the exterior is hydrophilic.

This protein is very similar to structures that have been designed *de novo*. The approach that people have taken for *de novo* design was to look first at the proportion of the various amino acids found in helices in proteins of known three-dimensional structures. They found that certain amino acids are more likely than others to occur in helices. They have used those statistical properties to design molecules of about this size that fold into a four-helix bundle. These molecules were designed without using any direct sequence homology to known proteins. They found that they could design proteins that folded according to design to make compact structures.

Going beyond the design of a structure without function, John Stewart and colleagues of the University of Colorado placed certain catalytic residues the alpha helix bundle to *mimic* proteolytic enzymes. They placed the Residues dues at specific sites m the molecule chosen to be close enough spatially to each other to catalyze a hydrolysis reaction, and they did, in fact, obtain enzymatic activity.

Thus, some design work has been done and successful predictions have been made. Several key features have been essential to these successes. The researchers had a detailed knowledge of the structures that they were trying to mimic, and had identified the key features of these structures. They had a clear idea of the specific modifications that they wanted to make. The structures were small enough for current methodologies to enable the researchers to predict what the effects would be of the modifications that they intended to introduce.

Life is not very simple. Not all *alpha helices* in proteins are organized into a four-helix bundle. One of the other common structural motifs for alpha helices in proteins is seen in *hemoglobin*. Instead of the helices being oriented parallel to each other, they cross each other. The result is a structural motif with a very different shape from the four-helix bundle, even though the number of helices involved in each motif is similar.

Such differences are one reason why it is very difficult to predict from the amino acid sequence of a protein what will be the folding pattern of that protein. You can predict a high probability that a given sequence will contain a substantial proportion of alpha helices, but the way in which the helices associate to form a structural motif will depend both upon the specific properties of the helices (determined by the specific amino acid sequence of each helix) and upon the way in which the helices interact with other portions of the protein molecule. Structural motifs like the four-helix bundle and the *hemoglobin helix motif* can be termed a *structural domain*. A domain is a group of interacting secondary structure elements that can be viewed as one structural unit.

A number of protein structural domains have been identified. The beta strands form a barrel-like structure, with the beta strands rep resenting the staves of the barrel. Accordingly, these structures are called *beta barrels*. A larger beta barrel structure is found in *superoxide dismutase*. This barrel is a less twisted barrel than the one in the previous example. Here, the beta strands are more parallel to each other. Like the previous examples of alpha helical domains, this protein also has metal atoms associated with it, but it has very little helical structure.

Again, life is not so simple as to have all domains consist of either all alpha helices or all beta strands. The beta structure tends to be *hydrophobic* (not liking water) and inside the molecule. The helices tend to be hydrophobic along one surface and *hydrophilic* on the other surface, and are oriented with the hydrophobic side of the helices, essentially protecting the beta strands from water, and thus stabilizing the structure. As seen from the end view, the beta strands form a twisted sheet, which is typical of beta strands because of steric effects of the amino acids usually found in beta strands. It can also be seen that both sides of the beta sheet are covered by helices.

An illustration of the fact that proteins with entirely different functions can have similar structures is provided by carboxypeptidase. *Carboxypeptidase* is an enzyme that cleaves residues from the carboxy terminus of proteins. *Lactate dehydrogenase* is a *redox protein* (an enzyme that catalyzes the simultaneous oxidation of one substrate and the reduction of another substrate). Both enzymes have a similar supersecondary structural motif formed from a twisted beta sheet covered by alpha helices. It can be seen that proteins can be organized into families according to the presence of well-characterized, super secondary structural motifs.

Bacteriochlorophyll protein is an example of a *membrane protein*. It is almost all beta structure, with the beta structure being in contact with the membrane. This protein harvests light. Someone asked if photons could be used to transfer information to control *nano machines*. The chromophores of this protein absorb light, so this protein could be used to receive information by way of photons.

An example of a more complicated protein is *triose phosphate isomerase*. It is similar to the previous examples in that it consists of beta structure surrounded by alpha helices to protect the barrel from solvent. Panel B shows this beta barrel (which is more hydrophobic on both sides than the previous examples) to be different in shape from those examples.

Lessons for Molecular Nanotechnology

These many examples demonstrate that design criteria can be applied to building fairly large molecules. Biological molecules afford examples of many defined structural types that could be incorporated into designs. It is probably not possible to perform a *de novo* prediction of the complete structure of large, complicated molecules. However, such approaches are feasible with smaller molecules, given the help provided by the knowledge base of known protein structures.

Practical Considerations for Biological Production of Proteins

In general, using mammalian cells to produce proteins means a substantial time lag between completing the design and producing a molecule of known purity. A number of problems are associated with using mammalian cells. One problem is that most proteins produced by mammalian cells have some carbohydrate side chains attached to the protein. Carbohydrate side chains are often variable in structure, so that the protein obtained may not be of defined structure (that is, not all molecules of the protein will have identical carbohydrate side chains attached). This is a significant problem for precise characterization of the protein produced by mammalian cells.

If bacterial cells are used to produce protein, the protein will not have any carbohydrate attached. However, some carbohydrate side chains are often necessary for the protein to be soluble. Therefore, the proteins are often produced by bacteria as insoluble "*inclusion bodies*" so that purification can be difficult. Sometimes it is not possible to produce a particular protein in bacteria.

In general, it takes two to three months from the time you decide to produce a specific mutation in a protein to the time that you can produce very small amounts of a fairly pure, mutated protein. That sounds relatively gloomy. However, you can make many proteins in parallel. As an example, a group of five or six people was able to produce about 250 mutant proteins, obtain information about their reaction kinetics and other structural information, on relatively small amounts of material, over a period of about a year to a year and a half. That means that a fair number of possibilities can be tried without enormous resources.

These capabilities mean that, at least for small proteins, it is possible to design a protein and to make a number of structures related to the design to test the design principles. What turns out to be the limiting factor in this whole process is the time that is required to determine the three-dimensional structures of the proteins that are produced. Often it is very difficult to crystallize proteins. Even if a protein can be crystallized, it still takes a significant amount of time to determine the structure by *X-ray crystallography*.

What seems to be indicated to accelerate progress with protein molecules are methods for solving structures (at least at modest resolution) in days. What we need for proteins is comparable to the capabilities that exist with small molecules to run a synthetic reaction—place a sample into a *nuclear magnetic resonance* (NMR) spectrometer, determine whether the desired synthesis was achieved, and if not, to walk back into the laboratory for another attempt the same day. The development of analytical techniques to do something similar for protein structures will be a key event in pushing forward the use of biotechnology to develop *molecular manufacturing*.

20

TRANSPORT IN NANOSTRUCTURES

The ability to transport charge (*electric current*), spin (*magnetic current*), and energy (heat, sound) through nanostructures is central to both scientific studies and technological applications. The quantum nature of matter on the nanometer scale leads to new transport phenomena and necessitates the development of new theoretical descriptions. In this chapter we focus on the *electrical transport properties* and treat some of the most important new quantum phenomena in nanostructures: *electron waves*, *tunneling*, *quantized conductance*, and *Coulomb blockade*.

NANOSTRUCTURES CONNECTED TO ELECTRON RESERVOIRS

A standard setup for measuring the transport properties of a nanostructure has been shown in figure. The electrons that flow through the nanostructure is supplied from macroscopic metal contacts. The energy and electron number of the large reservoirs are not affected by the presence of the nanostructure. It is thus natural to denote the metal contacts as electron reservoirs.

In the reservoirs the electrons behave more or less as classical particles. This is mainly due to the large volume V of the reservoirs, since a large V leads to a vanishingly small energy separation $\Delta\varepsilon$ between the quantum states in the reservoir, $\Delta\varepsilon \approx \hbar^2 / 2mV^{\frac{2}{3}}$. Thus given even the slightest perturbation the electrons will easily and frequently jump back and forth between quantum states whereby the distinct quantum nature is washed out. A given reservoir is characterized by its temperature T and its chemical potential μ, i.e. the free energy of the lastly added electron. The chemical potential can be changed in a controlled manner by applying a voltage V :

$$\mu = \mu_0 - eV. \quad ...(1)$$

The probability n_F that a given quantum state with energy å is occupied by an electron is determined by the *Fermi-Dirac distribution*:

$$n_F(\varepsilon, \mu_0, V, T) = \frac{1}{\exp\left[\frac{\varepsilon - \mu_0 + eV}{k_B T}\right] + 1}. \quad ...(2)$$

In the nanostructure between the reservoirs the electrons must however be described using quantum physics. This is due to the small transverse size w of the nanostructure which leads to appreciable *energy gaps* $\Delta\varepsilon$ given by $\Delta\varepsilon \approx h^2/2mw^2$. It now requires a substantial perturbation to force an electron to jump from one quantum state to the next. It is therefore very probable that a given electron stays in the same quantum state while passing through the nanostructure. The quantum properties are therefore not washed out and it can therefore be concluded that although the electrons are fed into the nanostructure as classical particles from a large reservoir the transport through the nanostructure must nevertheless be described in terms of the electron wave picture of quantum theory.

Current Density and Transmission of Electron Waves

We begin our analysis of electric conductance of nanostructures by studying simple 1D waves in piecewise constant potentials. We discuss the *electron waves*, then their current density, and finally the concept of transmission and reflection coefficients T and R.

Electron Waves in Constant Potentials in 1D

For an electron with energy ε moving in a given constant potential V the solution $\psi_\varepsilon(x)$ of the time-independent *Schrodinger equation*

$$-\frac{\hbar^2}{2m}\frac{\partial^2 \psi_\varepsilon(x)}{\partial x^2} + V\psi_\varepsilon(x) = E\psi_\varepsilon(x) \quad ...(3)$$

is given by

$$\psi_\varepsilon(x) = e^{\pm ikx}, \; k = \frac{1}{\hbar}\sqrt{2m(\varepsilon - V)}. \quad ...(4)$$

Note that $\varepsilon = \frac{\hbar^2 k^2}{2m} + V$. The two possible signs '+' and '–' correspond to waves travelling to the right and to the left, respectively. This can be seen by writing out the full time-dependent solutions $\psi_{+k}(x, t)$ and $\psi_{-k}(x, t)$:

$$\psi_{+k}(x,t) = e^{i(+kx - \frac{\varepsilon}{\hbar}t)}, \text{ (right-moving wave)}, \quad ...(5a)$$

$$\psi_{-k}(x,t) = e^{i(-kx - \frac{\varepsilon}{\hbar}t)}, \text{ (left-moving wave)}. \quad ...(5b)$$

In the following we suppress the *trivial time-dependence* $\exp(-i\frac{\varepsilon}{\hbar}t)$ and work only with the time-independent Schrodinger equation and the *time-independent solutions* $\psi_{\pm k}(x)$. But of course the concept of left- and right-moving waves remains the same:

$$\psi_{+k}(x) = e^{+ikx}, \text{ (right-moving wave)}, \qquad \text{...(6a)}$$

$$\psi_{-k}(x) = e^{-ikx}, \text{ (left-moving wave)}. \qquad \text{...(6b)}$$

Current Density J

The classical expression for current density is $J = nv = np/m$, n being the density of particles. In quantum theory the density of a single particle is $n = \psi^*\psi$, and a naive guess for the quantum expression for J would thus be $J = \psi^*\psi\ p/m$. However, the true result is slightly more complicated owing to the fact that in quantum theory $\hat{p}$ is a differential operator and not a number. Moreover, we need to ensure that J is real, we end up with

$$J(r) = \frac{1}{m}\mathrm{Re}[\psi^*(\hat{p}\psi)] = \frac{\hbar}{m}\mathrm{Im}[\psi^*(\nabla\psi)] \qquad \text{...(7)}$$

The electric current I passing through the yz plane in the x direction is given by

$$I = \int_{-\infty}^{\infty} dy \int_{-\infty}^{\infty} dz(-e)J_x = -\frac{e\hbar}{m}\int_{-\infty}^{\infty} dy \int_{-\infty}^{\infty} dz\, \mathrm{Im}\left[\psi^*\left(\frac{\partial\psi}{\partial x}\right)\right]. \qquad \text{...(8)}$$

In the case of simple product wavefunctions described in terms of some quantum numbers k in the x direction and α in the yz directions,

$$\psi_{k\alpha}(r) = X_k(x)\ \phi_\alpha(y, z), \qquad \text{...(9)}$$

we find that the current $I_{k\alpha}$ carried by the quantum state $\psi_{k\alpha}$ is

$$I_{k\alpha} = -\frac{e\hbar}{m}\mathrm{Im}\left[X_k(x)^*\frac{\partial}{\partial x}(X_k(x))\right]. \qquad \text{...(10)}$$

Transmission and Reflection Coefficients T and R in 1D

We now focus on 1D transport which is relevant if there is complete translation invariance in the yz plane or if there is only one

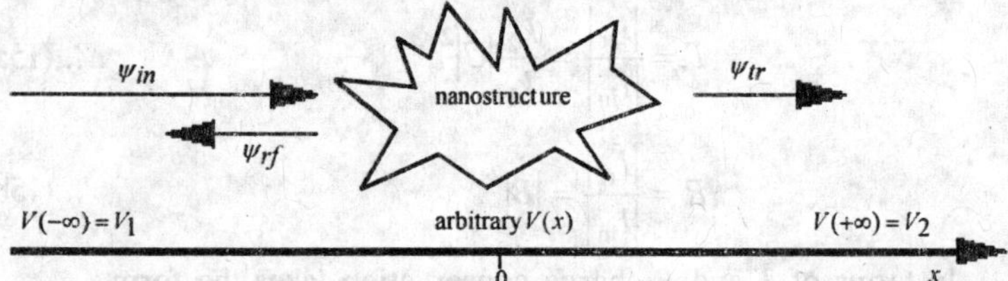

Fig. 20.1. Transport through a nanostructure formulated as an electron-wave scattering problem.

quantum state $\phi_\alpha(y, z)$ available for the yz part of the electron motion. The generic 1D scattering setup is shown in Fig. 20.1. We imagine that far to the left we have a constant potential V_1 and similarly another constant potential V_2 far to the right. Consider an incoming wave $\psi_{in}(x)$ at energy $\varepsilon > V_1, V_2$ from the far left. According to Eq. (4) we have

$$\psi_{in}(x) = \frac{1}{\sqrt{L}} e^{ik_1 x}, \quad k_1 = \frac{1}{\hbar}\sqrt{2m(\varepsilon - V_1)}, \qquad \text{...(11)}$$

where we now have inserted the *normalization factor* $1/\sqrt{L}$ taking the length L of the system into account. When this wave hits the scattering potential two things can happen: either it is reflected with some amplitude B back to the left in the state $\psi_{rf}(x)$ given by

$$\psi_{rf}(x) = \frac{B}{\sqrt{L}} e^{-ik_1 x}, \quad k_1 = \frac{1}{\hbar}\sqrt{2m(\varepsilon - V_1)}, \qquad \text{...(12)}$$

or it is transmitted with some amplitude C to the right in the state $\psi_{tr}(x)$ given by

$$\psi_{tr}(x) = \frac{C}{\sqrt{L}} e^{ik_2 x}, \quad k_2 = \frac{1}{\hbar}\sqrt{2m(\varepsilon - V_2)}. \qquad \text{...(13)}$$

Each of these states carries a current. From Eq. (10) we obtain

$$I_{in} = \frac{e\hbar}{mL}\text{Im}[ik_1] \qquad = -\frac{e\hbar}{mL}k_1, \qquad \text{...(14a)}$$

$$I_{rf} = -\frac{e\hbar}{mL}\text{Im}[ik_1]|B|^2 = -\frac{e\hbar}{mL}k_1|B|^2, \qquad \text{...(14b)}$$

$$I_{tr} = -\frac{e\hbar}{mL}\text{Im}[ik_2]|C|^2 = -\frac{e\hbar}{mL}k_2|C|^2. \qquad \text{...(14c)}$$

In steady-state charge conservation requires all of the incoming current to be either reflected or transmitted. Charge is not accumulated anywhere. It is therefore natural to define transmission and reflection coefficients T and R as follows:

$$T = \frac{|I_{tr}|}{|I_{in}|} = \frac{k_2}{k_1}|C|^2, \qquad \text{...(15a)}$$

$$R = \frac{|I_{rf}|}{|I_{in}|} = |B|^2. \qquad \text{...(15b)}$$

In terms of T and R charge conservation takes the form

$$T + R = 1. \qquad \text{...(16)}$$

ELECTRON WAVES AND THE SIMPLE POTENTIAL STEP

As the first example of treating transport with electron waves we treat the simple 1D potential step depicted in Fig. 20.2 and given by

$$V(x) = \begin{cases} 0, & \text{for } x < 0, \\ V, & \text{for } x > 0. \end{cases} \qquad \text{...(17)}$$

Consider first the situation shown in Fig. 20.2(a) where the energy ε of the electron is larger than the potential step V. Classically, we would expect full transmission when ε > V, but this is not so in quantum physics. Following the transmission and reflection coefficient we let a wave come in from the left and scatter on the potential step. The wave function takes the form

$$\psi_\varepsilon(x) = \begin{cases} e^{ik_1x} + Be^{-ik_1x},\ k_1 = \frac{1}{\hbar}\sqrt{2m\varepsilon} & \text{for } x < 0, \\ Ce^{ik_2x}, \quad k_2 = \frac{1}{\hbar}\sqrt{2m(\varepsilon - V} & \text{for} > 0. \end{cases} \qquad \text{...(18)}$$

The coefficients B and C are found by demanding continuity of $\psi_\varepsilon(x)$ and $\psi'_\varepsilon(x)$ at $x = 0$:

$$\psi_\varepsilon(0^-) = \psi_\varepsilon(0^+) \Rightarrow 1 + B = C, \qquad \text{...(19a)}$$

$$\psi'_\varepsilon(0^-) = \psi'_\varepsilon(0^+) \Rightarrow ik_1(1 - B) = ik_2C. \qquad \text{...(19b)}$$

From this follows

$$B = \frac{k_1 - k_2}{k_1 + k_2}, \text{ and } C = \frac{2k_1}{k_1 + k_2}. \qquad \text{...(20)}$$

From Eqs. (15a) and (15b) we then calculate the transmission and reflection coefficient T and R:

$$T = \frac{k_2}{k_1}|C|^2 = \frac{4k_1k_2}{(k_1 + k_2)^2}, \qquad \text{...(21a)}$$

$$R = |B|^2 = \frac{(k_1 - k_2)^2}{(k_1 + k_2)^2}. \qquad \text{...(21b)}$$

We note that $T + R = 1$ as required by the charge conservation Eq. (16). But we also note that even though $E > V$ we have $T < 1$, i.e. less than full transmission. Only in the limit of high energies ε > V, where $k_2/k_1 \to 1$, do we obtain $T \to 1$.

Consider then the situation of Fig. 20.2(b) where the energy ε of the electron is less than the potential step V. Classically, we would expect full reflection when ε < V, and this we also obtain in quantum physics. The wave function now takes the form

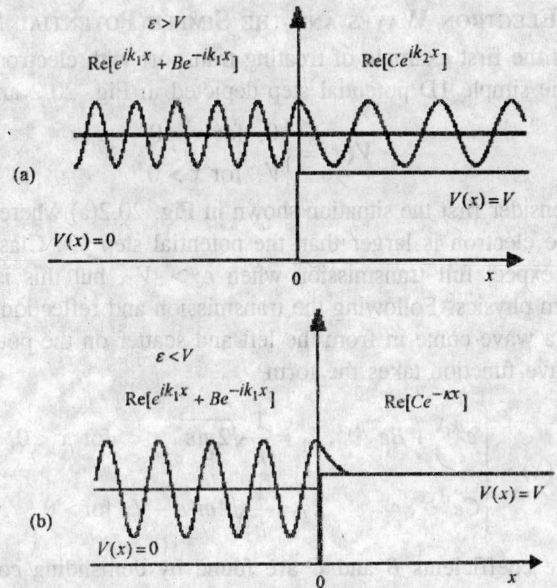

Fig. 20.2. A potential step of size V. (a) An electron wave with energy ε > V, and (b) with energy ε < V.

$$\psi_\varepsilon(x) = \begin{cases} e^{ik_1x} + Be^{-ik_1x}, & k_1 = \frac{1}{\hbar}\sqrt{2m\varepsilon} \quad \text{for } x < 0, \\ Ce^{-kx}, & k = \frac{1}{\hbar}\sqrt{2m(V-\varepsilon)} \quad \text{for } x > 0. \end{cases} \qquad ...(22)$$

Note how k_2 now is written as $i\kappa$ due to the sign change under the square root. As before the coefficients B and C are found by demanding continuity of $\psi_\varepsilon(x)$ and $\psi'_\varepsilon(x)$ at $x = 0$:

$$\psi_\varepsilon(0^-) = \psi_\varepsilon(0^+) \Rightarrow 1 + B = C, \qquad ...(23)$$

$$\psi'_\varepsilon(0^-) = \psi'_\varepsilon(0^+) \Rightarrow 1k_1(1 - B) = -\kappa C. \qquad ...(24)$$

From this follows

$$B = \frac{k_1 - i\kappa}{k_1 + i\kappa}, \text{ and } C = -\frac{2k_i}{k_1 + i\kappa}. \qquad ...(25)$$

Since in this case $k_2 = i\kappa$ we find from Eq. (14c) that $I_{tr} \propto$ Im $[ik_2]$ = Im$[-\kappa]$ = 0 and consequently

$$T = \frac{|I_{tr}|}{|I_{in}|} = 0. \qquad ...(26)$$

For the reflection coefficient we can use Eq. (15b):

$$R = |B|^2 = \frac{|k_1 - i\kappa|^2}{|k_1 + i\kappa|^2} = 1. \quad \text{...(27)}$$

We note that again $T + R = 1$ as required by charge conservation. We further note that as in the classical case we have total reflection when $\varepsilon < V$. Finally, we see that in spite of the total reflection there is in fact a finite probability to find the electron inside the potential barrier, a probability that is decaying exponentially with the distance from the potential step. This is a manifestation of the so-called quantum mechanical tunneling effect, an effect we study further in the following section.

Tunneling Through a Potential Barrier

Consider a rectangular potential barrier as the one sketched in Fig. 20.3. The barrier is defined by

$$V(x) = \begin{cases} 0, & \text{for } x < -a, \\ V, & \text{for } -a < x < a, \\ 0, & \text{for } x > a. \end{cases} \quad \text{...(28)}$$

We focus our attention on the general form of an eigenstate $\psi_\varepsilon(x)$ at the sub-barrier energy $\varepsilon < V$. As a straightforward generalization of Eq. (22) we write

$$\psi_\varepsilon(x) = \begin{cases} Ae^{ik_1x} + Be^{-ik_1x}, \quad k_1 = \frac{1}{\hbar}\sqrt{2m\varepsilon} & \text{for } x < -a, \\ Ce^{-\kappa x} + De^{\kappa x}, \quad \kappa = \frac{1}{\hbar}\sqrt{2m(V-\varepsilon)} & \text{for } -a < x < a, \\ Fe^{ik_1x} + Ge^{-ik_1x}, \quad k_1 = \frac{1}{\hbar}\sqrt{2m\varepsilon} & \text{for } a < x. \end{cases} \quad \text{...(29)}$$

In analogy with Eqs. (23) and (24) we demand continuity of $\psi_\varepsilon(x)$ and $\psi'_\varepsilon(x)$ at $x = -a$. We write the two resulting equations as one matrix equation:

$$\begin{pmatrix} e^{-ik_1a} & e^{ik_1a} \\ ik_1e^{-ik_1a} & -ik_1e^{ik_1a} \end{pmatrix}\begin{pmatrix} A \\ B \end{pmatrix} = \begin{pmatrix} e^{\kappa a} & e^{-\kappa a} \\ -\kappa e^{\kappa a} & \kappa e^{-\kappa a} \end{pmatrix}\begin{pmatrix} C \\ D \end{pmatrix}. \quad \text{...(30)}$$

We then multiply to the left with the inverse of the left matrix and get

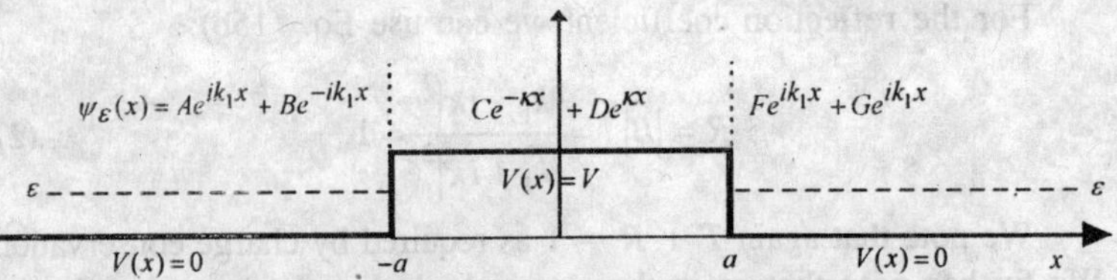

Fig. 20.3. A rectangular potential barrier of height V and width 2a.

$$\begin{pmatrix} A \\ B \end{pmatrix} = \frac{1}{2}\begin{pmatrix} \left(1+\frac{i\kappa}{k_1}\right)e^{\kappa a+ik_1a} & \left(1-\frac{i\kappa}{k_1}\right)e^{-\kappa a+ik_1a} \\ \left(1-\frac{i\kappa}{k_1}\right)e^{\kappa a-ik_1a} & \left(1+\frac{i\kappa}{k_1}\right)e^{-\kappa a-ik_1a} \end{pmatrix}\begin{pmatrix} C \\ D \end{pmatrix} \quad \text{...(31)}$$

In the same way continuity of $\psi_\varepsilon(x)$ and $\psi'_\varepsilon(x)$ at the other potential edge $x = a$ leads to a matrix equation for (C,D) and (F,G):

$$\begin{pmatrix} C \\ D \end{pmatrix} = \frac{1}{2}\begin{pmatrix} \left(1-\frac{ik_1}{\kappa}\right)e^{\kappa a+ik_1a} & \left(1+\frac{ik_1}{\kappa}\right)e^{-\kappa a+ik_1a} \\ \left(1+\frac{ik_1}{\kappa}\right)e^{\kappa a-ik_1a} & \left(1-\frac{ik_1}{\kappa}\right)e^{-\kappa a-ik_1a} \end{pmatrix}\begin{pmatrix} F \\ G \end{pmatrix}. \quad \text{...32}$$

Finally, by inserting Eq. (32) into Eq. (31) and multiplying the two matrices we obtain a relation between the wavefunction coefficients to the left and to the right of the barrier:

$$\begin{pmatrix} A \\ B \end{pmatrix} = \begin{pmatrix} \left(\cosh 2\kappa a + \frac{i\gamma}{2}\sinh 2\kappa a\right)e^{i2k_1a} & \frac{i\eta}{2}\sinh 2\kappa a \\ -\frac{i\eta}{2}\sinh 2\kappa a & \left(\cosh 2\kappa a - \frac{i\gamma}{2}\sinh 2\kappa a\right)e^{-i2k_1a} \end{pmatrix}\begin{pmatrix} F \\ G \end{pmatrix}, \quad \text{...(33)}$$

where we have introduced the abbreviations

$$\gamma = \frac{\kappa}{k_1} - \frac{k_1}{\kappa}, \text{ and } \eta = \frac{\kappa}{k_1} + \frac{k_1}{\kappa}. \quad \text{...(34)}$$

From Eq. (33) we are able to calculate the transmission and reflection coefficients T and R. We simply let Fig. 20.3 correspond to the scattering set-up sketched in Fig. 20.1, let a wave of amplitude $A = 1$ come in from the left and obtain a reflected and a transmitted wave of amplitude B for $x < -a$ and F for $x > a$, respectively. In this situation no wave comes in from the right towards the barrier, i.e. $G = 0$. With $A = 1$ and $G = 0$ Eq. (33) leads to

$$1 = F\left(\cosh 2\kappa a + \frac{i\gamma}{2}\sinh 2\kappa a\right)e^{i2k_1a}, \quad \text{...(35a)}$$

$$B = -F\frac{i\eta}{2}\sinh 2\kappa a, \quad \text{...(35b)}$$

and therefore

$$F=\frac{e^{-i2k_1a}}{\cosh(2\kappa a)+\frac{i\gamma}{2}\sinh(2\kappa a)}, \qquad \text{...(36a)}$$

$$B=\frac{-\frac{i\eta}{2}\sinh(2\kappa a)\,e^{-i2k_1a}}{\cosh(2\kappa a)+\frac{i\gamma}{2}\sinh(2\kappa a)}. \qquad \text{...(36b)}$$

From this and using Eqs. (15a) and (15b) with $k_2 = k_1$ and with F instead of C we obtain the transmission and reflection coefficients:

$$T=|F|^2=\frac{1}{\cosh^2(2\kappa a)+\frac{\gamma^2}{4}\sinh^2(2\kappa a)}=\frac{1}{1+\frac{\eta^2}{4}\sinh^2(2\kappa a)} \qquad \text{...(37a)}$$

$$R=|B|^2=\frac{\frac{\eta^2}{4}\sinh^2(2\kappa a)}{\cosh^2(2\kappa a)+\frac{\gamma^2}{4}\sinh^2(2\kappa a)}=\frac{\frac{\eta^2}{4}\sinh^2(2\kappa a)}{1+\frac{\eta^2}{4}\sinh^2(2\kappa a)}. \qquad \text{...(37b)}$$

We see that the charge conservation condition $R + T = 1$ is fulfilled.

Transmission Below the Barrier

The expression Eq. (36a) for the transmission amplitude F, combined with $G = 0$ and the matrix equation Eq. (32), gives us a direct way to calculate the amplitudes C and D in the barrier region. We thus have expressions for all amplitudes, and hence by Eq. (29) we can calculate the wavefunction $\psi_\varepsilon(x)$. It is clearly demonstrated how a tiny portion of the incoming wave tunnels through the barrier and continue its propagation on the other side of the barrier. Note the exponential decay of the amplitude in the barrier region. An explicit expression for T can be derived by first obtaining η^2 from Eqs. (29) and (34),

$$\eta^2=\left(\frac{k_1}{\kappa}+\frac{\kappa}{k_1}\right)^2=\frac{k_1^2}{\kappa^2}+\frac{\kappa^2}{k_1^2}+2=\frac{\varepsilon}{V-\varepsilon}+\frac{V-\varepsilon}{\varepsilon}+2=\frac{V^2}{\varepsilon(V-\varepsilon)} \qquad \text{...(38)}$$

so that

$$T(\varepsilon)=\left(1+\frac{V^2}{4\varepsilon(V-\varepsilon)}\sinh^2(2\kappa a)\right)^{-1}, \quad \text{valid for } \varepsilon < V. \qquad \text{...(39)}$$

Eq. (39) gives the transmission coefficient $T(\varepsilon)$ for all electron energies $\varepsilon < V$. In the limit of a high, $V > \varepsilon$, and wide, $\kappa a > 1$, barrier we can simplify the expression. Since $\sinh(x)=\frac{1}{2}[\exp(x)-\exp(-x)]\to\frac{1}{2}\exp(x)$, for $x\to\infty$, we find

$$T(\varepsilon)\approx 16\frac{\varepsilon}{V}e^{-4\kappa a}\approx 16\frac{\varepsilon}{V}e^{-4\sqrt{V/E_a}}, \quad \text{for } \varepsilon < V \text{ and } 1 < \kappa a. \qquad \text{...(40)}$$

In the second quasi-equality we have used the abbreviation $E_a = \hbar^2/2ma^2$. The expression reveals that the transmission coefficient decays exponentially with increasing barrier thickness a and with the square root of increasing barrier height, $\sqrt{V}$.

This result is truely remarkable. It tells us that if we take the mathematical formulation of quantum theory at face value, then we are forced to accept that particles can tunnel through potential barriers with a non-zero probability, a process that classical physics would rule out completely. It is an experimental fact that tunneling processes do occur in Nature. Let us mention a few examples of phenomena where tunneling is essential:

1. The α-decay of heavy nuclei is explained in terms of α-particles that tunnel out from the attractive nuclear force potential through the Coulomb-potential barrier.
2. The ionization of atoms in external electric fields weaker than the breakdown field is explained as electrons tunneling out through the Coulomb potential barrier to the region of lower potential created by the external field.
3. Modern, high precision voltage standards are based on the Josephson effect where pairs of electrons tunnel from one superconductor into another through an insulating oxide layer.
4. The scanning tunneling microscope, a key instrument in nanotechnolgy, is based on single-electrons tunneling from electron orbitals of the sample into the metal tip of the probe.

Transmission Above the Barrier

To finish the discussion of the rectangular potential barrier we now turn to the transmission for energies higher than the barrier, $\varepsilon > V$. In this situation the wavefunction $\psi_\varepsilon(x)$ from Eq. (29) becomes

$$\psi_\varepsilon(x) = \begin{cases} Ae^{ik_1x} + Be^{-ik_1x}, & k_1 = \frac{1}{\hbar}\sqrt{2m\varepsilon} \quad \text{for } x < -a, \\ Ce^{ik_2x} + De^{-ik_2x}, & k_2 = \frac{1}{\hbar}\sqrt{2m(\varepsilon - V} \quad \text{for } -a < x < a, \\ Fe^{ik_1x} + Ge^{-ik_1x}, & k_1 = \frac{1}{\hbar}\sqrt{2m\varepsilon} \quad \text{for } a < x. \end{cases} \quad ...(41)$$

The exponential terms $\exp(\mp\kappa x)$ are now oscillatory $\exp(\pm ik_2x)$, i.e. $k_2 = i\kappa$. We could now go through the same calculations for above-barrier transmission as for below-barrier transmission, i.e. match the wavefunction and its derivative at $x = \pm a$, to find the transmission

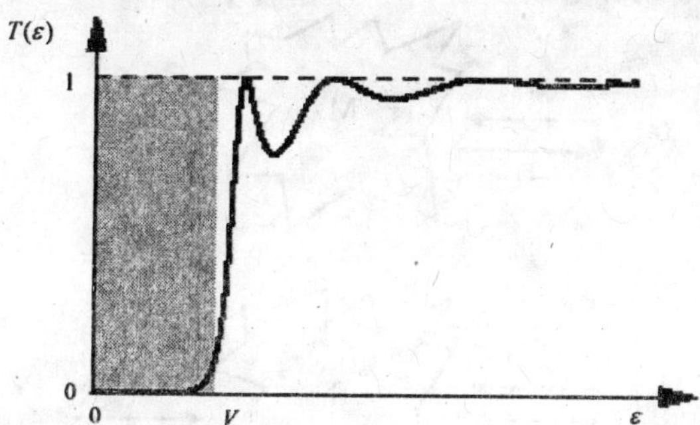

Fig. 20.4. The transmission coefficient $T(\varepsilon)$ for a rectangular potential barrier of height V and width $a=\sqrt{10\hbar^2/2mV}$.

coefficient $T(\varepsilon)$. Instead we will simply use the substitution $\kappa \to -ik_2$ in Eq. (39) and the identity $\sinh(ix) = \frac{1}{2}[\exp(ix)-\exp(-ix)] = i \sin(x)$ to obtain

$$T(\varepsilon)=\left(1+\frac{V^2}{4\varepsilon(\varepsilon-V)}\sin^2(2k_2a)\right)^{-1}, \text{ valid for } \varepsilon > V. \quad ...(42)$$

Complete Transmission Function T(ε)

Taken together Eqs. (39) and (42) give the complete transmission function $T(\varepsilon)$ for any positive value of ε. A specific graph of $T(\varepsilon)$ is shown in Fig. 20.4. The exponential suppression of the transmission is clearly seen in the gray barrier region $\varepsilon < V$.

As ε crosses over V and the above-barrier region is entered, the transmission starts to rise steaply. We note that $T(\varepsilon) \to 1$ for $\varepsilon \to \infty$, i.e. full transmission as expected for $\varepsilon > V$. We also note that for values of ε only slightly larger than V we generally do not obtain $T(\varepsilon) = 1$; there is above-barrier reflection. However, when the resonance condition $2k_2a = n\pi$, where $n = 1,2,3, \ldots$, is fulfilled peaks reaching full transmission appear.

The graph of $T(\varepsilon)$ demonstrates that one can encounter highly non-trivial transmission functions in quantum physics. The resonance structure can be expected each time sharp potential drops or well-defined quantum states, e.g. molecular orbitals, are present in the nanostructure. In other cases the transmission coefficient rises more monotonically from zero to unity.

Transfer and Scattering Matrices

The matrix equations Eqs. (31) and (32) are examples of the so-called *transfer matrix* method. In 1D one can collect the wave

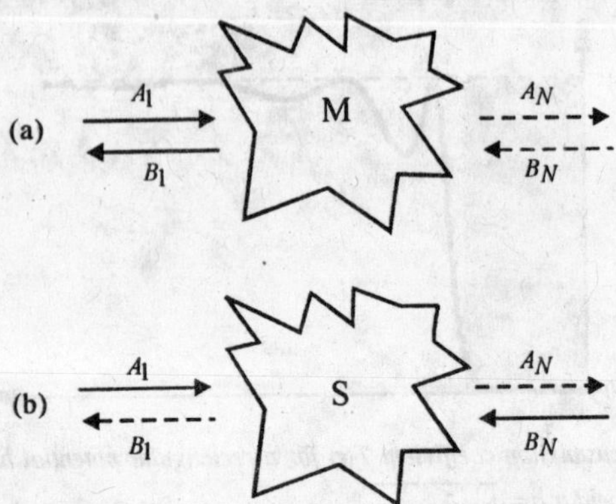

Fig. 20.5. Two possible ways to connect the wave amplitude A_1, B_1 A_N, and B_N. *(a) In 1D the transfer matrix M is often used to connect the amplitudes to the left, A_1 and B_1, with those to the right A_N and B_N. (b) In more general cases it is more convenient to use the scattering matrix S that connects the incoming amplitudes A_1 and B_1 with the outgoing ones B_1 and A_N.*

amplitudes to left, say (A_{N-1}, B_{N-1}), and those to the right, say (A_N, B_N), and then connect them by a matrix, the transfer matrix $\mathbf{M}_{N-1,N}$,

$$\begin{pmatrix} A_{N-1} \\ B_{N-1} \end{pmatrix} = M_{N-1,N} \begin{pmatrix} A_N \\ B_N \end{pmatrix}. \qquad \text{...(43)}$$

For a 1D system consisting of N falt potential regions, one can find the connection between the amplitudes of the first and the last region simply by successively apply the transfer matrices between neighbouring regions. The result is

$$\begin{pmatrix} A_1 \\ B_1 \end{pmatrix} = M_{1,2} M_{2,3} \cdots M_{N-2,N-1} M_{N-1,N} \begin{pmatrix} A_N \\ B_N \end{pmatrix} = M \begin{pmatrix} A_N \\ B_N \end{pmatrix}, \qquad \text{...(44)}$$

where M is the total transfer matrix of the system. A useful application of this formula is the following algorithm for calculating the transmission coefficient for any arbitrarily shaped potential. The given potential is approximated to any desired accuracy by N piecewise constant potential regions. We then follow Fig. 20.1 and connect a wave of the type $(1, B_1)$ to the left with $(A_N, 0)$ to the right,

$$\begin{pmatrix} 1 \\ B_1 \end{pmatrix} = M \begin{pmatrix} A_N \\ 0 \end{pmatrix}. \qquad \text{...(45)}$$

According to Eq. (15a) T is, in terms of the 1,1–component M_{11} of the transfer matrix,

$$T = \frac{k_N}{k_1}|A_N|^2 = \frac{k_N}{k_1}\frac{1}{M_{11}}. \qquad \text{...(46)}$$

For completeness we end this section by noting that in the most general scattering geometries, e.g. in higher dimensions than 1, it is often more convenient to work with the so-called *scattering matrix*, or simply *S-matrix*, instead of the transfer matrix. The scattering matrix relates the outgoing amplitudes (A_N, B_1) with the incoming amplitudes (A_1, B_N),

$$\begin{pmatrix} A_N \\ B_1 \end{pmatrix} = S\begin{pmatrix} A_1 \\ B_N \end{pmatrix}. \qquad \text{...(47)}$$

The S-matrix is useful because it is more directly related to fundamental concepts like conservation of charge and time-reversal symmetry. And in contrast to the transfer matrix the scattering matrix can always be defined in any scattering set-up. The scattering matrix is also fundamental in formulating conductance properties of quantum systems such as nanostructures. This is the topic of the following section.

Conductance and Scattering Matrix Formalism

We now return to the problem of finding the conductance $G=I/V$ of a generic nanostructure. We write the eigenstates of the electrons in the straight part of the sample between the nanostructure and the electron reservoirs in the form given by Eq. (9)

$$\psi_{k\alpha}(r) = \frac{1}{\sqrt{L}} e^{ikx} \phi_\alpha(y,z). \qquad \text{...(48)}$$

Electron Channels

For a given quantum number α there are many possible quantum states, namely two (spin up and spin down) for each k. One therefore talks about the electron channel α. The total energy of a state in a given channel α is denoted $\varepsilon_{k\alpha}$. It consists of two contributions: one denoted ε_α is the quantum energy of the transverse wavefunction $\phi_\alpha(y, z)$, while the other is the usual *kinetic energy* $\varepsilon_k = \hbar^2k^2/2m$ associated with the motion in the x direction. From a simple energy consideration we derive an expression for k:

$$\varepsilon_{k\alpha} = \varepsilon_k + \varepsilon_\alpha \Rightarrow k = \frac{1}{\hbar}\sqrt{2m(\varepsilon_{k\alpha} - \varepsilon_\alpha)}. \qquad \text{...(49)}$$

Note that the lowest possible energy for electrons in channel α is ε_α obtained when filling the state $\psi_{0\alpha}$.

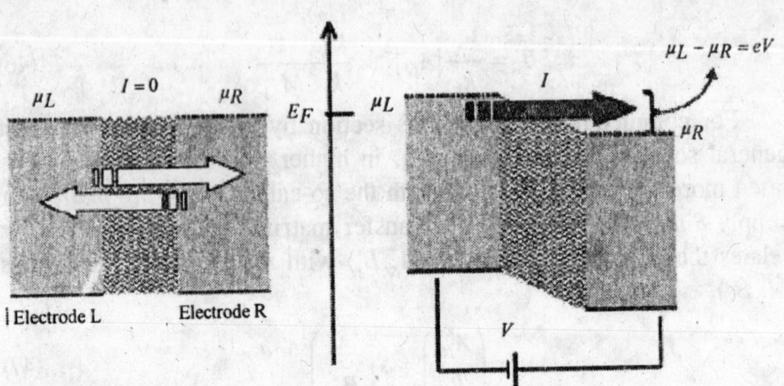

Fig. 20.6. Electrical currents through a nanostructure at zero temperature with injection of electrons from a left and a right electron reservoir.

Current, Reservoirs, and Electron Channels

We now analyze the current flow sketched in Fig. 20.6 starting with the current $I_{k\alpha}$ carried by an electron in state $\psi_{k\alpha}$ in channel α. It is given by Eq. (10), with a factor of 2 for spin inserted,

$$I_{k\alpha} = -\frac{2e\hbar}{Lm}\text{Im}\left[e^{-ikx}\frac{\partial}{\partial x}(e^{ikx})\right] = -\frac{2e\hbar k}{Lm} = -2\frac{e}{L}v_{k\alpha}, \quad ...(50)$$

where $v_{k\alpha}$ is the velocity of an electron in the state $\psi_{k\alpha}$.

It is now easy to derive an expression for the current $I_{L\to R}$ running from the left reservoir L into the right reservoir R through the nanostructure. We consider the energy $\varepsilon = \varepsilon_{k\alpha}$ and multiply three factors: (1) the Fermi-Dirac probability $n\frac{L}{F}(\varepsilon)$ that the right-moving state $\psi_{k\alpha}$ is occupied from the left reservoir; (2) the current $I_{k\alpha}$ carried by that state; and (3) the transmission probability $T\alpha(\varepsilon)$ of an electron at energy ε making it through to the reservoir to the right. Then we sum over all possible electron channels α and wavenumbers k:

$$I_{L\to R} = \sum_{\alpha}\sum_{k}\left[n\tfrac{L}{F}(\varepsilon)\right]\times\left[-2\tfrac{e}{L}v_{k\alpha}\right]\times\left[T_\alpha(\varepsilon)\right] \quad ...(51)$$

First we convert the k-sum into a k-integral which in 1D reads $\Sigma_k \to (L/2\pi)\int dk$. Then we convert the k-integral to an ε-integral: $\int dk \to \int d\varepsilon \frac{\partial k}{\partial \varepsilon}$. But we know that $\frac{\partial k}{\partial \varepsilon} = \frac{1}{\hbar v}$. So collecting all this yields $\Sigma_k \to (L/2\pi)\int d\varepsilon \frac{1}{\hbar v}$ and Eq. (51) becomes

$$I_{L\to R} = -\frac{2e}{h}\sum_{\alpha}\int_{-\infty}^{\infty} d\varepsilon\, n\tfrac{L}{F}(\varepsilon)\, T_\alpha(\varepsilon). \quad ...(52)$$

Note how all references to geometry, i.e. L, and velocity have vanished from the expression.

The current $I_{R\to L}$ flowing from the right to the left is obtained in a similar way,

$$I_{R\to L} = -\frac{2e}{h}\sum_{\alpha}\int_{-\infty}^{\infty} d\varepsilon\, n_F^R(\varepsilon)\, T_\alpha(\varepsilon). \qquad ...(53)$$

The total current I flowing in the system is therefore

$$I = I_{L\to R} - I_{R\to L} = -\frac{2e}{h}\sum_{\alpha}\int_{-\infty}^{\infty} d\varepsilon\, T_\alpha(\varepsilon)\left[n_F^L(\varepsilon) - n_F^R(\varepsilon)\right] \qquad ...54)$$

Conductance Formula for Nanostructures

We now have a relatively simple expression for the current, so to obtain the conductance we need to find the voltage dependence. In our set-up the *voltage* V is applied to the left reservoir. If we restrict ourselves to consider only small voltages (the linear response limit) it follows from Eq. (2) that

$$n_F^L(\varepsilon) - n_F^R(\varepsilon) = n_F(\varepsilon, \mu_0 - eV) - n_F(e, \mu_0) \approx \left.\frac{n_F}{\partial\mu}\right|_{\mu_0}$$

$$(-eV) = -\left(\frac{n_F}{\partial\varepsilon}\right)(-eV) \qquad ...(55)$$

With this Eq. (54) becomes

$$I = \frac{2e^2}{h}\sum_{\alpha}\int_{-\infty}^{\infty} d\varepsilon\, T_\alpha(\varepsilon)\left(-\frac{n_F}{\partial\varepsilon}\right)V. \qquad ...(56)$$

Finally, we obtain the temperature dependent conductance $G(T)$,

$$G(T) = \frac{I}{V} = \frac{2e^2}{h}\sum_{\alpha}\int_{-\infty}^{\infty} d\varepsilon\, T_\alpha(\varepsilon)\left(\frac{n_F}{\partial\varepsilon}\right) \qquad ...(57)$$

In the limit of very low temperature (temperatures around 1 K are routinely obtained in the lab cooling with liquid helium) a stated already that $-\frac{n_F}{\partial\varepsilon} = \delta(\varepsilon_F - \varepsilon)$. Inserting this in Eq. (56) yields the zero temperature conductance

$$G(T=0) = \frac{2e^2}{h}\sum_{\alpha} T_\alpha(\varepsilon_F). \qquad ...(58)$$

Eq. (57) is the conductance formula for nanostructures. We derived it by considering scattering states, i.e. the scattering matrix formalism. We note that the integral is dimensionless, so that means that the prefactor $G_0 = e^2/h$, depending entirely on universal constants, is some kind of conductance quantum, a natural unit for measuring conductance,

$$G_0 = \frac{e^2}{h} = 3.87404614 \times 10^{-5}\, S,$$

$$R_0 = \frac{1}{G_0} = \frac{h}{e^2} = 2.58128056 \times 10^4\, k\Omega. \qquad ...(59)$$

The zero temperature conductance Eq. (58) is simply the conductance quantum $\frac{2e^2}{h}$ times the sum of the transmission coefficients for each channel α evaluated at the Fermi energy set by the metal reservoirs. All information about the nanostructure lies in the transmission coefficient $T_\alpha(\varepsilon)$.

In the following section we shall see a spectacular consequence of the scattering wave nature of the conductance of nanostructures.

QUANTIZED CONDUCTANCE

The conductance quantum is directly observable in the beautiful experiment based on the quantum point contact. In the following we give a simple explanation of the observed quantization of the conductance. The quantum point contact is fabricated on a GaAs-GaAlAs heterostructure. As explained already all conduction electrons of this structure are bound to move at the interface between GaAs and GaAlAs, all having the same wavefunction $\zeta_0(z)$ in the z direction (perpendicular to the interface). By etching techniques a narrow wire is created in the x direction. The width w of the wire in the y direction varies with position, $w(x)$, being of the order 100 nm at the narrowest point. The actual width can be controlled during the experiment by changing the gate voltage V_g on the side electrodes.

Let us model the potential $V_x(y)$ of the wire at each position x as a simple potential box stretching $w/2$ on both sides of the x axis:

$$V_x(y) = \begin{cases} 0, & \text{for } -\frac{1}{2}w(x) < y < \frac{1}{2}w(x), \\ \infty, & \text{for } |y| > \frac{1}{2}w(x). \end{cases} \qquad ...(60)$$

We further imagine that the change of width as we move along the x axis is so slow that a simple product wavefunction fitting to the width at any given position is a good solution to the *Schrodinger equation*. This approximation which effectively neglects the derivatives $w'(x)$ and $w''(x)$ is known of the adiabatic approximation. The explicit wavefunctions are

$$\psi_{\varepsilon n0}(x,y,z) = \frac{1}{\sqrt{L}} e^{ik_{\varepsilon n}(x)x} \sqrt{\frac{2}{w(x)}} \sin\left(n\pi\left[\frac{y}{w(x)} + \frac{1}{2}\right]\right)\zeta_0(z). \qquad ...(61)$$

The wavefunction is fully characterized by the energy ε, the number n of standing half-waves in the wire and the number 0 reminding us that all electrons have the same wave $\zeta_0(z)$ in the z direction. The index α used prior is here a double index $\alpha = n0$. The total energy ε of the state $\psi_{\varepsilon n0}$ can be written as

$$\varepsilon = \varepsilon_k + \varepsilon_n(x) + \varepsilon_z = \frac{\hbar^2 k_{\varepsilon n}(x)^2}{2m} + \frac{\pi^2 \hbar^2}{2m} \frac{n^2}{w(x)^2}. \quad ...(62)$$

Here we have chosen to put the zero point of the energy scale at the value ε_z of the quantum energy of $\zeta_0(z)$, i.e. $\varepsilon_z = 0$. We can interpret Eq. (62) as the energy equation for a particle moving along the x-axis with a position dependent kinetic energy $\varepsilon_k = \frac{1}{2m}\hbar^2 k_{\varepsilon n}(x)^2$ and a position dependent potential energy $\varepsilon_n(x)$ given by

$$\varepsilon_n(x) = \frac{\pi^2 \hbar^2}{2m} \frac{n^2}{w(x)^2}. \quad ...(63)$$

As the particle moves towards the narrowest point of the point contact it must convert more and more of its kinetic energy in the forward x direction into potential energy (really kinetic energy in the transverse y direction). To find out whether the particle actually has enough kinetic energy to pass through it is of course now highly relevant to know what is the maximal potential energy ε_n^{max} along the wire. The smallest width of the wire is denoted w_{min}, then according to Eq. (63) we have

$$\varepsilon_n^{max} = \frac{\pi^2 \hbar^2}{2m} \frac{n^2}{w_{min}^2}. \quad ...(64)$$

The x-dependent wavenumber $k_{\varepsilon n}(x)$ is given by

$$k_{\varepsilon n}(x) = \sqrt{\frac{2m}{\hbar^2}(\varepsilon - \varepsilon_n(x))}. \quad ...(65)$$

We see that in our simple model we obtain full transmission if $k_{\varepsilon n}(x)$ remains real throughout the passage, i.e. if $\varepsilon < \varepsilon_n^{max}$. If however we arrive at $\varepsilon > \varepsilon_n^{max}$ then $k_{\varepsilon n}(x)$ becomes imaginary corresponding to tunneling through the effective potential barrier $\varepsilon_n(x)$. As a rough model for the transmission coefficient T_n of channel n, i.e. all electrons having a transverse wave consisting of n half-waves, we simply take

$$T_n(\varepsilon) = \begin{cases} 1, & \text{for } \varepsilon > \varepsilon_n^{max} \\ 0, & \text{for } \varepsilon < \varepsilon_n^{max}. \end{cases} \quad ...(66)$$

With the transmission coefficient Eq. (66) inserted into Eq. (58) we obtain an explanation of the zero temperature conductance of the quantum point contact. The resulting formula is namely

$$G(T=0,V_g)=\frac{2e^2}{h}\sum_n T_{n,V_g}(\varepsilon_F), \qquad ...(67)$$

where we explicitly write that the conductance must depend on the gate voltage V_g that controls the actual minimal width of the wire, i.e. $w^{min} = w^{min}(V_g)$. If we initially have a large negative V_g the channel will be pinched off entirely. No electrons can pass and $G = 0$. As we gradually increase V_g the wire gets wider and wider. At some point electrons in channel $n = 1$ will be able to pass, since they have the lowest threshold potential $\varepsilon_{n=1}^{max}$, and the conductance jumps from 0 up to $2e^2/h$. As V_g is increased further we reach a point where also the electrons in the $n = 2$ channel can pass, and the conductance increases from $2e^2/h$ to $4e^2/h$. This process goes on as we increase V_g more and more, and we thus trace out a conductance curve like the lowest one ($T = 1.3$ K).

This result is a dramatic deviation from classical physics. The conductance does not increase linearly with decreasing width, but instead it increases in discrete steps of height $2e^2/h$, the conductance quantum. Another deviation from Ohm's law is achieved if one places two quantum point contacts in series close to one another. If we assume that point contact 1 is more narrow than point contact 2, then electron waves can only pass through the second if they are able to pass through the first. So the total conductance G of the system is entirely determined by one of the point contacts, $G = \min\{G_1, G_2\} = G_1$. In terms of the electrical resistance R we have $R = \max\{R_1, R_2\}$. But this is radically different form the usual classical formula $R = R_1 + R_2$. We only recover the classical result if the distance between the point contacts is so large that the quantum waves are destroyed in the region between them, and the whole electron wave picture thus breaks down.

21

DECISION MAKING MACHINES

Computers have emerged from back rooms and laboratories to help with writing, calculating, and play in homes and offices. These machines do simple, repetitive tasks, but machines still in the laboratory do much more. *Artificial intelligence* researchers say that computers can be made smart, and fewer and fewer people disagree. To understand our future, we must see whether artificial intelligence is as impossible as flying to the Moon.s

Thinking machines need not resemble human beings in shape, purpose, or mental skills. Indeed, some artificial intelligence systems will show few traits of the intelligent liberal arts graduate, but will instead serve only as powerful engines of design. Nonetheless, understanding how human minds evolved from mindless matter will shed light on bow machines can be made to think. Minds, like other forms of order, evolved through variation and selection.

Minds act. One need not embrace *Skinnerian behaviourism* to see the importance of behaviour, including the internal behaviour called *thinking*. RNA replicating in test tubes shows how the idea of purpose can apply (as a kind of shorthand) to utterly mindless molecules. They lack nerves and muscles, but they have evolved to "*behave*" in ways that promote their replication. Variation and selection have shaped each molecule's simple behaviour, which remains fixed for its whole "life." Individual RNA molecules don't adapt, but bacteria do. Competition has favoured bacteria that adapt to change, for example by adjusting their mix of digestive enzymes to suit the food available. Yet these mechanisms of adaptation are themselves fixed: food molecules trip genetic switches as cold air trips a thermostat.

Bacteria

Some bacteria also use a primitive form of *trial-and-error* guidance. Bacteria of this sort tend to swim in straight lines, and have just enough "*memory*" to know whether conditions are improving or worsening as they go. If they sense that conditions are improving, they keep going straight. If they sense that conditions are getting worse, they stop, tumble, and head off in a random, generally different, direction. They test directions, and favour the good directions by discarding the bad. And because this makes them wander toward concentrations of food molecules, they have prospered.

Flatworm

Flatworm lack brains, yet show the faculty of true learning. They can learn to choose the correct path in a simple T-maze. They try turning left and turning right, and gradually select the behaviour—or form the habit—which produces the better result. This is selection of behaviour by its consequences, which behaviourist psychologists call "the *Law of Effect*." The evolving genes of worm species have produced worm individuals with evolving behaviour.

Law of Effect

Still, worms trained to run mazes (even Skinner's pigeons, trained to peck when a light flashes green) show no sign of the reflective thought we associate with mind. Organisms adapting only though the simple *Law of Effect* learn only by *trial and error*, by varying and selecting actual behaviour—they don't think ahead and decide. Yet natural selection often favoured organisms that could think, and thinking is not magical. As Daniel Dennett of Tufts University points out, evolved genes can equip animal brains with internal models of bow the world works (somewhat like the models in computer-aided engineering systems). The animals can then "*imagine*" various actions and consequences, avoiding actions which "*seem*" dangerous and carrying out actions which "*seem*" safe and profitable. By testing ideas against these internal models, they can save the effort and risk of testing actions in the external world.

Dennett further points out that the *Law of Effect* can reshape the models themselves. As genes can provide for evolving behaviour, so they can provide for evolving mental models. Flexible organisms can vary their models and pay more attention to the versions that prove better guides to action. We all know what it is to try things, and learn which work. Models need not be instinctive; they can evolve in the course of a single life.

Speechless Animals

Speechless animals, however, seldom pass on their new insights. These vanish with the brain that first produced them, because learned mental models are not stamped into the genes. Yet even speechless animals can imitate each other, giving rise to memes and cultures. A female monkey in Japan invented a way to use water to separate grain from sand; others quickly learned to do the same. In human cultures, with their language and pictures, valuable new models of how the world works can outlast their creators and spread worldwide.

Mind

On a still higher level, a mind (and "*mind*" is by now a fitting name) can hold evolving standards for judging whether the parts of a model–the ideas of a worldview–seem reliable enough to guide action. The mind thus selects its own contents, including its selection rules. The rules of judgment that filter the contents of science evolved in this way.

Goals

As behaviour, models, and standards for knowledge evolve, so can *goals*. That which *brings* good, as judged by some more basic standard, eventually begins to *seem* good: it then becomes a goal in itself. Honesty pays, and becomes a valued principle of action. As thought and mental models guide action and further thought, we adopt clear thinking and accurate models as goals in themselves. Curiosity grows, and with it a love of knowledge for its own sake. The evolution of goals thus brings forth both science and ethics. As Charles Darwin wrote, '*the highest possible stage in moral culture is when we recognize that we ought to control our thoughts.*" We achieve this as well by *variation* and *selection*, by concentrating on thoughts of value and letting others slip from attention.

Marvin Minsky of the MIT *Artificial Intelligence Laboratory* views the mind as a sort of society, an evolving system of communicating, cooperating, competing agencies, each made up of yet simpler agents. He describes thinking and action in terms of the activity of these agencies. Some agencies can do little more than guide a hand to grasp a cup; others (vastly more elaborate) guide the speech system as it chooses words in a sticky situation. We aren't aware of directing our fingers to wrap around a cup *just so*. We delegate such tasks to competent agents and seldom notice unless they slip. We all feel conflicting impulses and speak unintended words; these are symptoms of discord among the agents of the mind. Our awareness of this is part of the self-regulating process by which our most general agencies manage the rest.

Memes

Memes may be seen as agents in the mind that are formed by teaching and imitation. To feel that two ideas conflict, you must have embodied both of them as agents in your mind—though one may be old, strong, and supported by allies, and the other a fresh idea-agent that may not survive its first battle. Because of our superficial self-awareness, we often wonder where an idea in our heads came from. Some people imagine that these thoughts and feelings come directly from agencies outside their own minds; they incline toward a belief in haunted heads.

Genius

In ancient Rome, people believed in "*genii,*" in good and evil spirits attending a person from cradle to grave, bringing good and ill luck. They attributed outstanding success to a special "*genius.*" And even now, people who fail to see how natural processes create novelty see "*genius*" as a form of magic. But in fact, evolving genes have made minds that expand their knowledge by varying idea patterns and selecting among them. With quick variation and effective selection, guided by knowledge borrowed from others, why shouldn't such minds show what we call genius? Seeing intelligence as a natural process makes the idea of intelligent machines less startling. It also suggests how they might work.

Machine Intelligence

One dictionary definition of "*machine*" is "Any system or device, such as an electronic computer, that performs or assists in the performance of a human task." But just how many human tasks will machines be able to perform? Calculation was once a mental skill beyond machines, the province of the intelligent and educated. Today, no one thinks of calling a pocket calculator an *artificial intelligence*; calculation now seems a "*merely*" mechanical procedure.

Still, the idea of building ordinary computers once was shocking. By the mid-1800s, though, Charles Babbage had built mechanical calculators and part of a programmable mechanical computer; how ever, he ran into difficulties of finance and construction. One Dr. Young helped not at all: he argued that it would be cheaper to invest the money and use the interest to pay human calculators. Nor did the

British Astronomer Royal, Sir George Airy—an entry in his diary states that "On September 15th Mr. Goulburn ... asked my opinion on the utility of Babbage's calculating machine ... I replied, entering fully into the matter, and giving my opinion that it was worthless."

Babbage's machine was ahead of its time—meaning that in building it, machinists were forced to advance the art of making precision parts. And in fact it would not have greatly exceeded the speed of a skilled human calculator—but it would have been more reliable and easier to improve.

The story of *computers* and *artificial intelligence* (known as *AI*) resembles that of flight in air and space. Until recently people dismissed both ideas as impossible—commonly meaning that they couldn't see how to do them, or would be upset if they could. And so far, *AI* has had no simple, clinching demonstration, no equivalent of a working airplane or a landing on the Moon. It has come a long way, but people keep changing their definitions of intelligence.

Press reports of "*giant electronic brains*" aside, few people called the first computers intelligent. Indeed, the very name "*computer*" suggests a mere arithmetic machine. Yet in 1956, at Dartmouth, during the world's first conference on artificial intelligence, researchers Alan Newell and Herbert Simon unveiled Logic Theorist, a program that proved theorems in symbolic logic.

In later years computer pro grams were playing chess and helping chemists determine molecular structures. Two medical programs, *CASNET* and *MYCIN* (the first dealing with internal medicine, the other with the diagnosis and treatment of infections), have performed impressively. According to the *Handbook of Art Intelligence*, they have been "rated, in experimental evaluations, as performing at human-expert levels in their respective domains." A program called *prospector* has located, in Washington state, a molybdenum deposit worth millions of dollars.

These so-called "*expert systems*" succeed only within strictly limited areas of competence, but they would have amazed the computer programmers of the early 1950s. Today, however, few people consider them to be *real* artificial intelligence: *AI* has been a moving target. The passage from *Business Week* quoted earlier only shows that computers can now be programmed with enough knowledge, and perform fancy enough tricks, that some people feel comfortable calling them intelligent. Years of seeing fictional robots and talking computers on television have at least made the idea of *AI* familiar.

The chief reason for declaring *AI* impossible has always been the notion that "*machines*" are intrinsically stupid, an idea that is now beginning to fade. Past machines have indeed been gross, clumsy things that did simple, brute-force work. But computers handle information, follow complex instructions, and can be instructed to change their own instructions, They can experiment and learn. They contain not gears and grease but traceries of wire and evanescent patterns of electrical energy. As Douglas Hofstadter urges (through a character in a dialogue about *AI*), "Why don't you let the word '*machine*' conjure up images of patterns of dancing light rather than of giant steam shovels?"

Cocktail-party critics confronted with the idea of artificial intelligence often point to the stupidity of present computers, as if this proved something about the future. (A future machine may wonder whether such critics exhibited genuine thought.) Their objection is irrelevant—steam locomotives didn't fly, though they demonstrated mechanical principles later used in airplane engines. Likewise, the creeping worms of an eon ago showed no noticeable intelligence, yet our brains use neurons much like theirs.

Casual critics also avoid thinking seriously about *AI* by declaring that we can't possibly build machines smarter than ourselves. They forget what history shows. Our distant, speechless ancestors managed to bring forth entities of greater intelligence through genetic evolution without even thinking about it. But we are thinking about it, and the memes of technology evolve far more swiftly than the genes of biology. We can surely make machines with a more human-like ability to learn and organize knowledge.

There seems to be only one idea that could argue for the impossibility of making thought patterns dance in new forms of matter. This is the idea of *mental materialism*—the concept that mind is a special substance, a magical thinking-stuff somehow beyond imitation, duplication, or technological use.

Psychobiologists see no evidence for such a substance, and find no need for mental materialism to explain the mind. Because the complexity of the brain lies beyond the full grasp of human understanding, it seems complex enough to embody a mind. Indeed, if a single person could *fully* understand a brain, this would make the brain less complex than that person's mind. If all Earth's billions of people could cooperate in simply watching the activity of one human brain, each person would have to monitor tens of thousands of active synapses simultaneously—clearly an impossible task. For a person to try to understand the flickering patterns of the brain as a whole would be five billion times more absurd. Since our *brain's mechanism* so massively overwhelms our mind's ability to grasp it, that mechanism seems complex enough to embody the mind itself.

TURING'S TARGET

In a 1950 paper on machine intelligence, British mathematician Alan Turing wrote: "I believe that by the end of the century the use of words and general educated opinion will have altered so much that one will be able to speak of machines thinking without expecting to be contradicted." But this will depend on what we call thinking. Some say that only people can think, and that computers cannot be people; they then sit back and look smug.

But in his paper, Turing asked how we judge *human* intelligence, and suggested that we commonly judge people by the quality of their conversation. He then proposed what he called the *imitation game*—which everyone else now calls the Turing test. Imagine that you are in a room, able to communicate through a terminal with a person and a computer in two other rooms. You type messages; both the person and the computer can reply. Each tries to act human and intelligent. After a prolonged keyboard "*conversation*" with them—perhaps touching on literature, art, the weather, and how a mouth tastes in the morning—might be that you could not tell which was the person and which the machine. If a machine could converse this well on a regular basis, then Turing suggests that we should consider it genuinely intelligent. Further, we would have to acknowledge that it knew a great deal about human beings.

For most practical purposes, we need not ask "Can a machine have *self-awareness*—that is, *consciousness*?" Indeed, critics who declare that machines cannot be conscious never seem able to say quite what they mean by the term. Self-awareness evolved to guide thought and action, not merely to ornament our humanity. We must be aware of other people, and of their abilities and inclinations, to make plans that involve them. Likewise we must be aware of ourselves, and of our own abilities and inclinations, to make plans about ourselves. There is no special mystery in self awareness. What we call the self-reacts to impressions from the rest of the mind, orchestrating some of its activities; this makes it no more (and no less) than a special part of the interacting patterns of thought. The idea that the self is a pattern in a special mind substance (distinct from the mind substance of the brain) would explain nothing about awareness.

A machine attempting to pass the Turing test would, of course, claim to have *self-awareness*. Hard-core biochauvinists would simply say that it was lying or confused. So long as they refuse to say what they mean by consciousness, they can never be proved wrong. Nonetheless, whether called *conscious* or not, *intelligent machines* will still *act* intelligent, and it is their actions that will affect us. Perhaps they will someday shame the biochauvinists into silence by impassioned argument, aided by a brilliant public-relations campaign.

No machine can now pass the *Turing test*, and none is likely to do so soon. It seems wise to ask whether there is a good reason even to try: we may gain more from *AI* research guided by other goals.

Let us distinguish two sorts of artificial intelligence, though a system could show both kinds. The first is *technical AI*, adapted to deal with the physical world. Efforts in this field lead toward automated engineering and scientific inquiry. The second is *social AI*, adapted to deal with human minds. Efforts in this field lead toward machines able to pass the *Turing test*.

Researchers working on *social AI* systems will learn much about the human mind along the way, and their systems will doubtless have great practical value, since we all can profit from intelligent help and advice. But automated engineering based on *technical AI* will have a greater impact on the technology race, including the race toward molecular technology. And an advanced automated engineering system may be easier to develop than a *Turing-test passer*, which must not only possess knowledge and intelligence, but must mimic *human* knowledge and *human* intelligence—a special more difficult challenge.

As Turing asked, "May not machines carry out something which ought to be described as thinking but which is very different from what a man does?" Although some writers and politicians may refuse to recognize machine intelligence until they are confronted with a talkative machine able to pass the Turing test, many engineers will recognize intelligence in other forms.

Engines of Design

We are well on the way to *automated engineering*. Knowledge engineers have marketed expert systems that help people to deal with practical problems. Programmers have created computer-aided design systems that embody knowledge about shapes and motion, stress and strain, electronic circuits, heat flow, and how machine tools shape metal. Designers use these systems to augment their mental models, speeding the evolution of yet unbuilt designs. Together, designers and computers form intelligent, semiartificial systems.

Engineers can use a wide variety of computer systems to aid their work. At one end of the spectrum, they use computer screens simply as drawing boards. Farther along, they use systems able to describe parts in three-dimensions and calculate their response to heat, stress, current, and so on. Some systems also know about computer-controlled manufacturing equipment, letting engineers make simulated tests of instructions that will later direct computer-controlled machines to make real parts. But the far end of the spectrum of systems involves using computers not just to record and test designs, but to generate them.

Programmers have developed their most impressive tools for use in the computer business itself. Software for chip design is an example. *Integrated circuit chips* now contain many thousands of transistors and wires. Designers once had to work for many months to design a circuit to do a given job, and to lay out its many parts across the surface of the chip. Today they can often delegate this task to a so-called "*silicon compiler*." Given a specification of a chip's function, these software systems can produce a detailed design—ready for manufacture—with little or no human help.

All these systems rely entirely on human knowledge, labouriously gathered and coded. The most flexible automated design systems today can fiddle with a proposed design to seek improvements, but they learn nothing applicable to the next design. But EURISKO is different. Developed by Professor Douglas Lenat and others at Stanford University, EURISKO is designed to explore new areas of knowledge. It is guided by heuristics—pieces of knowledge that suggest plausible actions to follow or implausible ones to avoid; in effect, various rules of thumb. It uses heuristics to suggest topics to work on, and further heuristics to suggest what approaches to try and how to judge the results. Other heuristics look for patterns in results, propose new heuristics, and rate the value of both new and old heuristics. In this way EURISKO evolves better behaviours, better internal models, and better rules for selecting among internal models. Lenat himself describes the variation and selection of heuristics and concepts in the system in terms of "mutation" and "selection," and suggests a social, cultural metaphor for understanding their interaction.

Since heuristics evolve and compete in EURISKO, it makes sense to expect parasites to appear—as indeed many have. One machine-generated heuristic, for example, rose to the highest possible value rating by claiming to have been a co-discoverer of every valuable new conjecture. Professor Lenat has worked closely with EURISKO, improving its mental immune system by giving it heuristics for shed ding parasites and avoiding stupid lines of reasoning.

EURISKO has been used to explore elementary mathematics, programming, biological evolution, games, three-dimensional integrated circuit design, oil spill cleanup, plumbing, and (of course) heuristics. In some fields it has startled its designers with novel ideas, including new electronic devices for the emerging technology of three-dimensional integrated circuits.

The results of a tournament illustrate the power of a human/*AI* team. Traveller TCS is a futuristic naval war game, played in accordance with two hundred pages of rules specifying design, cost, and performance constraints for the fleet ("*TCS*" stands for "*Trillion Credit Squadron*"). Professor Lenat gave EURISKO these rules, a set of starting heuristics, and a program to simulate a battle between two fleets. He reports that "it then designed fleet after fleet, using the simulator as the '*natural selection*' mechanism as it '*evolved*' better and better fleet designs." The program would run all night, designing, testing, and drawing lessons from the results. In the morning Lenat would cull the designs and help it along. He credits about 60 percent of the results to himself, and about 40 percent to EURISKO.

Lenat and EURISKO entered the 1981 national Traveller TCS tournament with a strange fleet. The other contestants laughed at it, then lost to it. The Lenat/EURISKO fleet won every round, emerging as the national champion. As Lenat notes, "This win is made more significant by the fact that no one connected with the program had ever played this game before the tournament, or seen it played, and there were no practice rounds."

In 1982 the competition sponsors changed the rules. Lenat and EURISKO entered a very different fleet. Other contestants again laughed at it, then lost. Lenat and EURISKO again won the national championship.

In 1983 the competition sponsors told Lenat that if he entered and won again, the Competition would be canceled. Lenat bowed out.

EURISKO and other *AI* programs show that computers need not be limited to boring, repetitive work if they are given the right sort of programming. They can explore possibilities and turn up novel ideas that surprise their creators. EURISKO has shortcomings, yet it points the way to a style of partnership in which an AI system and a human expert both contribute knowledge and creativity to a design process.

In coming years, similar systems will transform engineering. Engineers will work in a creative partnership with their machines, using software derived from current computer-aided design systems for doing simulations, and using evolving, EURISKO-like systems to suggest designs to simulate. The engineer will sit at a screen to type in goals for the design process and draw sketches of proposed designs. The system will respond by refining the designs, testing them, and displaying proposed alternatives, with explanations, graphs, and diagrams. The engineer will then make further suggestions and changes, or respond with a new task, until an entire system of hard ware has been designed and simulated.

As such automated engineering systems improve, they will do more and more of the work faster and faster. More and more often, the engineer will simply propose goals and then sort among good solutions proposed by the machine. Less and less often will the engineer have to select parts, materials, and configurations. Gradually engineers will be able to propose more general goals and expect good solutions to appear as a matter of course. Just as EURISKO ran for hours evolving fleets with a *Traveller TCS simulator*, automated engineering systems will someday work steadily to evolve passenger jets having maximum safety and economy—or to evolve military jets and missiles best able to control the skies.

Just as EURISKO has invented electronic devices, future automated engineering systems will invent molecular machines and molecular electronic devices, aided by software for molecular simulations. Such advances in automated engineering will magnify the design-ahead phenomenon described earlier. Thus automated engineering will not only speed the assembler breakthrough, it will increase the leap that follows.

Eventually software systems will be able to create bold new designs without human help. Will most people call such systems intelligent? It doesn't really matter.

AI Race

Companies and governments worldwide support *AI* work because it promises commercial and military advantages. The United States has many university artificial intelligence laboratories and a host of new companies with names like Machine Intelligence Corporation,

Thinking Machines Corporation, Teknowledge, and Cognitive Systems Incorporated. In October of 1981 the Japanese Ministry of Trade and Industry announced a ten-year, $850 million program to develop advanced AI hardware and software. With this, Japanese researchers plan to develop systems able to perform a billion logical inferences per second. In the fall of 1984 the Moscow Academy of Science announced a similar, five-year, $100 million effort. In October of 1983 the U.S. Department of Defense announced a five-year, $600 million Strategic Computing Program; they seek machines able to see, reason, understand speech, and help manage battles. As Paul Wallich reports in the *IEEE Spectrum*, "Artificial intelligence is considered by most people to be a cornerstone of next-generation computer technology; all the efforts in different countries accord it a prominent place in their list of goals.'

Advanced AI will emerge step by step, and each step will pay off in knowledge and increased ability. As with molecular technology (and many other technologies), attempts to stop advances in one city, county, or country will at most let others take the lead. A miraculous success in stopping visible *AI* work *everywhere* would at most delay it and, as computers grow cheaper, let it mature in secret, beyond public scrutiny. Only a world state of immense power and stability could truly stop *AI* research everywhere and forever—a "*solution*" of bloodcurdling danger, in light of past abuses of merely national power. Advanced AI seems inevitable. If we hope to form a realistic view of the future, we cannot ignore it.

In a sense, *artificial intelligence* will be the ultimate tool because it will help us build all possible tools. *Advanced AI* systems could maneuver people out of existence, or they could help us build a new and better world. Aggressors could use them for conquest, or fore sighted defenders could use them to stabilize peace. They could even help us control AI itself. The hand that rocks the AI cradle may well rule the world.

As with assemblers, we will need foresight and careful strategy to use this new technology safely and well. The issues are complex and interwoven with everything from the details of molecular technology to employment and the economy to the philosophical basis of human rights. The most basic issues, though, involve what AI can do.

Are We Smart Enough?

Despite the example of the evolution of human beings, critics may still argue that our limited intelligence may somehow prevent us from programming genuinely intelligent machines. This argument seems weak, amounting to little more than a claim that because the critic can't see how to succeed, no one else will ever do better. Still, few would deny that programming computers to equal human abilities will indeed require fresh insights into human psychology. Though the programming path to AI seems open, our knowledge does not justify the sort of solid confidence that thoughtful engineers had (decades before Sputnik) in being able to reach the Moon with rockets, or that we have today in being able to build assemblers through protein design. Programming genuine artificial intelligence, though a form of engineering, will require new science. This places it beyond firm projection.

We need accurate foresight, though. People clinging to comforting doubts about AI seem likely to suffer from radically flawed images of the future. Fortunately, automated engineering escapes some of the burden of biochauvinist prejudice. Most people are less upset by the idea of machines designing machines than they are by the idea of true general-purpose *AI systems*. Besides, automated engineering has been shown to work; what remains is to extend it. Still, if more general systems are likely to emerge, we would be foolish to omit them from our calculations. Is there a way to sidestep the question of our ability to design intelligent programs?

In the 1950s, many *AI researchers* concentrated on simulating brain functions by simulating neurons. But researchers working on programs based on words and symbols made swifter progress, and the focus of AI work shifted accordingly. Nonetheless, the basic idea of neural simulation remains sound, and molecular technology will make it more practical. What is more, this approach seems guaranteed to work because it requires no fundamental new insights into the nature of thought.

Eventually, *neurobiologists* will use *virus-sized molecular machines* to study the structure and function of the brain, cell by cell and molecule by molecule where need be. Although *AI* researchers may gain useful insights about the organization of thought from the resulting advances in brain science, neural simulation can succeed without such insights. Compilers translate computer programs from one language to another without understanding how they work. Photocopies transfer patterns of words without reading them. Like wise, researchers will be able to copy the neural patterns of the brain to another medium without understanding their higher-level organization.

After learning how neurons work, engineers will be able to design and build analogous devices based on *advanced nano electronics* and *nanomachines*. These will interact like neurons, but will work faster. Neurons, though complex, do seem simple enough for a mind to understand and an engineer to imitate. Indeed, neurobiologists have learned much about their structure and function, even without molecular machinery to probe their workings.

With this knowledge, engineers will be able to build fast, capable AI systems, even without understanding the brain and without clever programming. They need only study the brain's neural structure and join 5 neurons to form the same functional pattern. If they make all the parts right—including the way they mesh to form the whole—then the whole, too, will be right. "*Neural*" activity will flow in the patterns we call thought, but faster, because all the parts will work faster.

Accelerating the Technology Race

Advanced AI systems seem possible and inevitable, but what effect will they have? No one can answer this in full, but one effect of automated engineering is clear: it will speed our advance toward the limits of the possible.

To understand our prospects, we need some idea of how fast *advanced AI systems* will think. Modern computers have only a tiny fraction of the brain's complexity, yet they can already run programs imitating significant aspects of human behaviour. They differ totally from the brain in their basic style of operation, though, so direct physical comparison is almost useless. The brain does a huge number of things at once, but fairly slowly; most modern computers do only one thing at a time, but with blinding speed.

Still, one can imagine *AI hardware* built to imitate a brain not only in function, but in structure. This might result from a neural-simulation approach, or from the evolution of *AI programs* to run on hard ware with a brain like style of organization. Either way, we can use analogies with the human brain to estimate a *minimum* speed for advanced assembler-built *AI systems*.

Neural synapses respond to signals in thousandths of a second; experimental electronic switches respond a hundred million times faster (and nano electronic switches will be faster yet). Neural signals travel at under one hundred meters per second; electronic signals travel a million times faster. This crude comparison of speeds suggests that brain like electronic devices will work about a million times faster than brains made of neurons (at a rate limited by the speed of electronic signals).

This estimate is crude, of course. A neural synapse is more complex than a switch; it can change its response to signals by changing its structure. Over time, synapses even form and disappear. These changes in the fibers and connections of the brain embody the long-term mental changes we call learning. They have stirred Professor Robert Jastrow of Dartmouth to describe the brain as an enchanted loom, weaving and reweaving its neural patterns throughout life.

To imagine a brain like device with comparable flexibility, picture its electronic circuits as surrounded by mechanical nanocomputers and assemblers, with one per synapse-equivalent "*switch*. "Just as the molecular machinery of a synapse responds to patterns of neural activity by modifying the synapse's structure, so the nanocomputers will respond to patterns of activity by directing the nanomachinery to modify the switch's structure. With the right programming, and with communication among the nanocomputers to simulate chemical signals, such a device should behave almost exactly like a brain.

Despite its complexity, the device will be compact. *Nanocomputers* will be smaller than *synapses*, and assembler-built wires will be thinner than the brain's *axons* and *dendrites*. Thin wires and small switches will make for compact circuits, and compact circuits will speed the flow of electronic patterns by shortening the distances signals must travel. It seems that a structure similar to the brain will fit in less than a cubic centimeter. Shorter signal paths will then join with faster transmission to yield a device over ten million times faster than a human brain.

Only cooling problems might limit such machines to slower average speeds. Imagine a conservative design, a million fold faster than a brain and dissipating a million fold more heat. The system consists of an assembler-built block of sapphire the size of a coffee mug, honeycombed with circuit-lined cooling channels. A high-pressure water pipe of equal diameter is bolted to its top, forcing cooling water through the channels to a similar drainpipe leaving the bottom. Hefty power cables and bundles of optical-fiber data channels trail from its sides.

The cables supply fifteen megawatts of electric power. The drainpipe carries the resulting heat away in a three-ton-per-minute flow of boiling-hot water. The optical fiber bundles carry as much data as a million television channels. They bear communications with other *AI* systems, with engineering simulators, and with assembler systems that build designs for final testing. Every ten seconds, the system gobbles almost two kilowatt-days of electric energy (now worth about a dollar). Every ten seconds, the system completes as much design work as a human engineer working eight hours a day for a year (now worth tens of thousands of dollars). In an hour, it completes the work of centuries. For all its activity, the system works in a silence broken only by the rush of cooling water.

This addresses the question of the sheer speed of thought, but what of its complexity? *AI development* seems unlikely to pause at the complexity of a single human mind. As John McCarthy of Stanford's *AI lab* points out, if we can place the equivalent of one human mind in a metal skull, we can place the equivalent of ten thousand cooperating minds in a building. (And a large modern power plant could supply power enough for each to think at least ten thousand times as fast as a person.) To the idea of *fast engineering intelligences*, add the idea of fast engineering teams.

Engineering AI systems will be slowed in their work by the need to perform experiments, but not so much as one might expect. Engineers today must perform many experiments because bulk technology is unruly. Who can say in advance exactly how a new alloy will behave when forged and then bent ten million times? Tiny cracks weaken metal, but details of processing determine their nature and effects.

Because assemblers will make objects to precise specifications, the unpredictabilities of bulk technology will be avoided. Designers (whether human or *AI*) will then experiment only when experimentation is faster or cheaper than calculation, or (more rarely) when basic knowledge is lacking.

AI systems with access to *nanomachines* will perform many experiments rapidly. They will design apparatus in seconds, and replicating assemblers will build it without the many delays (ordering special parts, shipping them, and so on) that plague projects today. Experimental apparatus on the scale of an assembler, nanocomputer, or living cell will take only minutes to build, and nanomanipulators will perform a million motions per second. Running a million ordinary experiments at once will be easy. Thus, despite delays for experimentation, automated engineering systems will move technology forward with stunning speed.

From past to future, then, the likely pattern of advancing ability looks something like this. Across eons of time, life moved forward in a long, slow advance, paced by genetic evolution. Minds with language picked up the pace, accelerated by the flexibility of memes. The invention of the methods of science and technology further accelerated advances by forcing memes to evolve faster. Growing wealth, education, and population—and better physical and intellectual tools—have continued this accelerating trend across our century.

The automation of engineering will speed the pace still more. Computer-aided design will improve, helping human engineers to generate and test ideas ever more quickly. Successors to EURISKO will shrink design times by suggesting designs and filling in the de tails of human innovations. At some point, full-fledged automated engineering systems will pull ahead on their own.

In parallel, molecular technology will develop and mature, aided by advances in automated engineering. Then assembler-built *AI systems* will bring still swifter automated engineering, evolving technological ideas at a pace set by systems a million times faster than a human brain. The rate of technological advance will then quicken to a great upward leap: in a brief time, many areas of technology will advance to the limits set by natural law. In those fields, advance will then halt on a lofty plateau of achievement.

This transformation is a dizzying prospect. Beyond it, if we survive, lies a world with replicating assemblers, able to make whatever they are told to make, without need for human labour. Beyond it, if we survive, lies a world with automated engineering systems able to direct assemblers to make devices near the limits of the possible, near the final limits of technical perfection.

Eventually, some *AI systems* will have both great technical ability and the social ability needed to understand human speech and wishes. If given charge of energy, materials, and assemblers, such a system might aptly be called a "*genie machine.*" What you ask for, it will produce. Arabian legend and universal common sense suggest that we take the dangers of such engines of creation very seriously indeed.

Decisive breakthroughs in technical and *social AI* will be years in arriving. As Marvin Minsky has said, "The modestly intelligent machines of the near future promise only to bring us the wealth and comfort of tireless, obedient, and inexpensive servants." Most systems now called "*AI*" do not think or learn; they are only a crude distillate of the skills of experts, preserved, packaged, and distributed for consultation.

But *genuine AI* will arrive. To leave it out of our expectations would be to live in a fantasy world. To expect *AI* is neither optimistic nor pessimistic: as always, the researcher's optimism is the technophobe's pessimism. If we do not prepare for their arrival, social *AI systems* could pose a grave threat: consider the damage done by the merely human intelligence of terrorists and demagogues. Likewise, technical *AI systems* could destabilize the world military balance, giving one side a sudden, massive lead. With proper preparation, however, artificial intelligence could help us build a future that works—for the Earth, for people, and for the advancement of intelligence in the universe.

Why discuss the dangers today? Because it is not too soon to start developing institution able to deal with such question. *Technical AI* is emerging today, and its every advance will speed the technology race. Artificial intelligence is but one of many powerful technologies we must learn to manage, each adding to a complex mixture of threats and opportunities.

22

MECHANICAL MAN

Back in the 1940s, the young Isaac Asimov took a staple of science fiction, the anthropoid robot, and made it his own. His series of *robot* stories and novels defined for the rest of the century; but particularly for the futuristic 1950s, what a robot should look like and what it could do. Physically, a *robot* was a machine made of steel in roughly the shape of a man. It had machinelike properties: it was strong, did not get tired, did not feel pain, and did not need to eat. The real *robots* of today are typically made of aluminum, plastic, or composites, as well as steel; but the ones of tomorrow will be gemstone balloons or banks of fog.

Mentally, the robot had a *positronic brain*. This was just glitzy science fiction verbiage. *Positrons* had recently been discovered, so the obvious science fictional step past electronic brains was positronic ones. In reality; as a particle of antimatter, a positron would be a bull in a china shop in any actual machine, annihilating itself and the first electron it ran into and releasing hard gamma radiation.

Skipping lightly over that, then, Asimov patterned his *first robot* after an animal. It could understand a limited amount of language but could not talk. It was not too bright, but was trustworthy like a dog. Later experience with *real machines* has shown us that talking is easy but understanding is difficult, something we have long suspected from our dealings with human beings. *Robots* today are much more prevalent than people may realize. Your car and your computer were likely partially made by robots. Industrial robots are hugely expensive machines that must operate in a carefully planned and controlled environment, because they have very limited senses and no common sense whatsoever. With nanotechnology that will change drastically. Indeed, it is already starting to change, as the precursor technologies such as *micromachines* begin to have their effect.

Existing robots are often stupider than insects. As computers increase in power, however robots will get smarter, be more able to operate in unstructured environments, and ultimately be able to do anything a human can. They will find increasing use, as costs come clown, in production, in service industries, and as domestic servants.

The physical body of a useful robot is easily within the capability of current technology. A hobbyist call make one capable of per forming a variety of household tasks for a few thousand dollars in parts. The sticking point is the controller—a computer that reads the sensors and tells the motors what to do—and the sensors. *Robots* are getting more common almost by the minute nowadays because the cost of computers, signal acquisition circuitry, video cameras, microphones, miniature gyros and accelerometers, pressure sensors, and the like are falling through the floor.

Once the sensors arc all there—and today they cost more than the bones and muscles—there remains the problem of programming the controller to interpret them and drive the actuators so as to achieve useful, coordinated activity. This is not a completely solved problem but you might be surprised at the progress that is been made in just the past decade.

The American Association for *Artificial intelligence* holds a "*robot challenge*" among other robot competitions at its annual conference. These are for real, autonomous robots, not remote control drones. The challenge requires a robot to "start at the entrance to the conference center, find the registration desk, register for the conference, perform volunteer duties as required, and report at a prescribed time in a conference hail to give a talk." All by itself. At the 2003 meeting, two robots, one from Carnegie Mellon and one from Washington University, completed the challenge.

With *nanotechnology*; *robots* could have a sensory nervous system as complex and sensitive as the human one. That means millions of sensors where current robots have to get by with less than one hundred. *Nanocomputers* will provide plenty of processing power to integrate and interpret what the sensors are sensing. Just as radical an advance will be available on the motor side. Instead of the ten or twenty motors that today's awkward robots must get by with, *nanotech robots* could contain hundreds, making them as supple and graceful as a person, or thousands, giving them the flexibility and fluidity of an octopus. A balloon made of thin sheets of diamond, the size and shape of a human arm (with internal partitions to maintain the shape), would need fabric just a couple of microns thick to be as strong as the arm. Make it 5 microns to allow for foldable structures, motors, local controllers, communication and power networks, and so forth. Don't stop with the arm—make a full human-sized balloon. That is your robot. United, the fabric occupies 10 cubic centimeters (a ballpoint pen), and weighs less than an ounce.

The *robot* might need ballast to stay on the ground in a stiff breeze, out otherwise it is physically the size and strength of a human, probably a lot more flexible and graceful, and tireless as long its fuel supply lasts. If folding up was not an issue, you could fill the balloon with fuel—it would still weigh less than a human—and run for a very long time indeed.

In the old science fiction novels, a *robot* who was not being used would stand in a niche in the wall, like a statue. In reality; you will hold out your hand, it will collapse to the size of a pen, and you will put it in your pocket. That is, if it does not simply vanish.

Utility Fog

The physical properties of matter, how hard or strong something is, whether it is a solid or liquid, depend on how its *molecules* react to forces that are applied to it. If they resist any force that tries to move them relative to each other, it a solid. If they allow very small forces to push them past each other but still try to maintain a fixed distance to their neighbours, it a liquid. If they try to push apart no matter how widely spread they already are, it is a gas.

Soft solids, like foam rubber, resist distorting forces weakly but with a force that increases with distance stretched. Hard ones like diamond resist with a force that increases enormously with even the tiniest microscopic change in shape. Metals resist up to a point and then flow like fluids. Modeling clay is similar, out the transition point is a much lower force.

Modeling clay and foam rubber, of course, are not simple materials. Their properties derive in part from structure well above the molecular scale. The same is true of wood, paper, bone, muscle, concrete, and indeed most common materials besides plastic. Even metals have grain structure that causes their properties to vary. Heat treatment that changes the grain structure can double the strength of the metal. If you can stand some graininess, you can often get a lot of flexibility in return. Look at your television screen: the picture is composed of dots that can change colour. The dots do not move, but the picture can appear to as the pattern of colour moves across the dots. If all the dots change colour at once, one picture can vanish and be replaced by another instantly.

The dots on your TV screen have a size, and a speed of changing, that is designed to he small and fast enough to maintain an illusion of a moving picture to your eyes. It generally works, even though you can see the individual dots if you look closely and see a flicker out of the corner of your eye. If you swap the 300,000 dots of the TV for the 130 million of a high-quality laser-printed page, the illusion becomes almost perfect.

With *nanotechnology*, we can make *robots* the size of those 1,200 to the inch dots on the laser-printed page. These robots can change properties so as to simulate the range of properties of ordinary matter, resisting a push or moving in the direction pushed, simply by running a different control program.

Take a box full of such *robots*. You now have the solid equivalent of the TV screen, where each three-dimension dot can take on properties that are not just colours but the physical characteristics of matter. Like pictures on the TV, tangible physical objects can appear, interact, and vanish with a broadcast command telling each robot which program to run. Unlike TV pictures, the virtual objects could pick up and early real ones, have weight, he hard, soft, springy, or even liquid. They could do real work in the real world.

Fill your entire house with the *robots*, instead of just a box, and you'll be able to have furniture, appliances robotic servants and even your clothing appear and disappear on command.

The stuff consisting of a mass of such robots is called *Utility Fog*. When it was invented it circa 1990, it was the first "*swarm robot*" concept with "*instant-on*" virtual objects. Since other writers, notably Ray Kurzweil, have opined that it should have a better name, "*Utility Fog*" cam from. First, the stuff fills the air like fog, and you walk around in it. It would look a lot like fog as well. As for the "*Utility*" part, remember the *Batman* TV series from the 1960s? Whenever Batman needed some gadget, lo and behold, there it was in his "utility belt." There seemed to be no end of what the belt could produce—and it was always right at hand. Living in *Utility Fog* might be a lot like being in a campy, tongue in-cheek TV show, or even in a cartoon. Cartoon characters are constantly reaching up and grabbing guns, hammers, anvils, suits of clothing, false beards, tables, chairs, signs dealing witty remarks, and the occasional steam locomotive out of thin air.

What about walking around it, though? Would not it be like trying to walk around in modeling clay? No, because the vast majority of the stuff would be running the "*simulate air*" program even though it is denser than air, about the same density as balsa wood, you would not feel it because it is powered and would actively get out of your way. There are lots of options for time specifics of the human—Fog interface, but one that seems reasonable is that it would maintain a bubble around your bead, like a spacesuit helmet.

"*Foglets*," the individual robots, on the inner surface of the bubble would be equipped with phase-controlled light emitters, so they could project a real-time hologram of what you would expect to see around you, instead of looking into a fogbank. Of course, the Fog would have a lot more information about what's out there than just its surface colour—it's touching everything in sight—so it could give you false-colour images revealing other properties or make objects appear to be translucent so you could see through them, or whatever you like.

Or it could display the appearance of an entirely different location and simulate the feel of any object there. Suppose I am standing in an empty cubicle—empty that is, except that it is filled with *Utility Fog*. There happens to be a high-speed data link from my cubicle to your living room. The Fog bubble around my head shows me a view as if I were sitting across from you, having afternoon tea. So far, no difference in principle from having a video camera and my watching a screen. But I decide to sit down, the Foglets around me getting out of the way to let me move, and the scat of my pants gets to the point where my eyes tell me your couch is.

The Foglets at that point are not simulating air, but couch. They feel like a couch. They hold me up like a couch. The Foglets in front of me are simulating the coffee table and I can put my teacup down on them a little too hard, hear a thump, and observe a splash. The teacup will work best if it is real and really there, and the tea itself *must* be real if I'm to drink it. That's telepresence—half of it, anyway. What happening in your real living room?

In the simplest case, a robot sits there and transmits what it feels only so that the *Fog* can produce the same sensations for me. This might be an ordinary robot, but it could also be a Fog robot—a mass of Fog the same size and shape as me. This would have the advantage that when I "left," the batch of Fog could be put to other uses. But the most interesting case would be if your living room Were completely full of Fog too.

Now, instead of having to walk into your living room in robotic form, I can suddenly appear there, as the Foglets in a certain area quit simulating air and start simulating me. I can do anything there that I could if I were there in person—Utility Fog is at least as strong as human flesh—pick up objects, open and close doors, drink tea. But I don't need to drink actual tea in your living room, I have my real teacup in my cubicle. So the teacup in your living room that I seem to be holding can be made of Fog, too, and the tea as well—since I am not there to really drink it.

Let's suppose that the chair you are sitting on is really there in your living room, but the coffee table is not. It is a virtual object formed by the Fog. Here we sit, apparently together, apparently in the same place, but we are actually in two different places. Some of the objects we are using are actually with you, and are being simulated where I am. Some are actually with me, and are simulated where you are. Some do not really exist at all, and are simulated both places.

The more of the common objects that are simulated both places, the closer you are to pure virtual reality.

We are talking, of course—that's what tea is for. The subject of robotics comes up. You ask your house for "*Asimov's Robot novels*" and a small bookcase appears in the air to your right. You reach out, grab, say; I, *Robot*, and hand it over. I summon my own bookcase in a similar manner and stick it in. Both bookcases disappear.

What actually happened? It was a file transfer. When I call up my bookcase next time, I'll find the book there, and I'll be able to open it and read it. It's no different, in principle, from dragging an icon across your computer screen from one folder to another.

Another form of telepresence is scale shifting. Suppose you want to build a *skyscraper* or *nanomachine*. In either case you are better off if the parts fit in your fingers' grasp easily and the whole business sits in your lap or on your workbench. With appropriately sized robotics at the other end, this is just; how it would look and feel. As in the living room case, the robot could be an outsized (or nanoscale) humanoid mechanism, hut in many cases it makes more sense for it to be *Utility Fog*, too.

Utility Fog would be nowhere near as good as special-purpose *nanoengineered machines* for doing any specific task. It would weigh about as much as balsa wood, about as strong as *high-density polyethylene* (HDPE) plastic, and be about as hard. But it could simulate weight, strength, and hardness well enough to fool your sense of touch. It would be perfectly strong enough to simulate furniture, but it could not simulate a knife hard enough to actually cut something. However, it would be an extremely facile manipulator of objects, so a batch of Fog could have a stock of its own microscopic tools, like knives, to do the cutting where the simulation demanded it.

The same is true for computation, communication, and anything involving heat or chemistry (like cooking). Power and control would be provided by special-purpose units that the *Fog* would early around like any other object. *Individual Foglets* would have simple onboard controllers and tiny fuel tanks, but there would be sand grain sized units every few millimeters, pea-sized ones every few inches, golf ball-sized ones every few feet.

Foglets themselves would have small, roughly spherical bodies with twelve arms coming out in all directions. They could be any size above about ten microns armtip to armtip; a twelve hundred dots- per-inch resolution, good enough to fool your sense of touch, calls for 20 microns. If resolution to fool senses was not important, the *robots* could even be of macroscopic size. Thus we don't really need nanotechnology to start designing and experimenting with Utility Fog robots and software. But we do need nanotechnology for its ability to manufacture high-tech items in astronomical quantitie^. A typical room, floor to ceiling, would contain literally quadrillions of Foglets.

You can reduce this considerably, down to a few hundred trillion or so, if you are willing to give up some of the nonphysical effects such as having objects appear instantly and hover in the air. What you do is have a thick Fog carpet instead of filling the air. Then, when you want a chair, for example, the Foglets have to move physically to flow up into the shape you want. (This is called *naive mode* and the other is called *Fog mode*). While not instant, *Fog flow* could easily be fast enough to form a chair in less than a second.

Fail-safe operations for the naive mode would differ from *Fog mode*. It would be OK to have naive-mode objects freeze into shape upon a loss of power or control signals, but in Fog mode, the air in particular would have to be able to get out of your way. If all backup power sources were exhausted, the air would retract and clump into marble or golf ball-sized spheres and leave the real air there. You'd find yourself standing in something like those play areas that are knee-deep in plastic balls that kids like so much. Both modes have advantages. In naive mode the control is considerably simpler. In Fog mode you can simulate a barbell that would be heavier than balsa wood—the air surreptitiously pulls down on the parts you are not touching. In naive mode you do not have to generate and run a holographic simulation of the surroundings. In Fog mode you can have objects appear and disappear and levitate real objects as well as virtual ones, including yourself.

Flying

How would you like to leap tail buildings in a single bound? Run faster than a locomotive—or pick one up? I am sorry, you would not be able to crush coal to diamonds in your hands, but hidden synthesizers could make plenty of diamonds for you.

Flying would be particularly fun. We can distinguish between virtual reality flying, where you feel like you are flying but are really sort of swimming in your cubicle, and really flying out in the open over real territory. The real form, of course, requires a lot of *Utility Fog* to cover the whole area. No magic—the reason you stay up in the air is that a stack of robots, albeit very tiny robots, holds you up.

Covering a vast area with a mile-thick layer of microscopic robots may seem extravagant, but I am sure that covering Manhattan with a layer of buildings would have seemed extravagant to any of its inhabitants in 1492. It seems likely that a layer of *Utility Fog* would be one of the most efficient forms for the city of the future. A ten-foot cube of *Fog* provides enough room for virtual reality of anything for one person, and the flying ability solves the transportation problem for physical meetings within the city. The *Fog* supplies all the infrastructure that a current-clay city has and much more.

There will likely be areas—large domes, sonic cities, property around houses—that are Fog-filled, but not everywhere. The advantages of filling your house with Fog we have already seen; objects can appear and disappear on command, you can accomplish most physical tasks simply by asking for them. *Globs of Fog* outside allow you the same luxuries in your yard, while preventing unwanted intruders—people, clogs, mosquitoes, or bacteria. *Flying* could also be useful in space. Aren't you already weightless in space? Yes, but you do not have control: you cannot stop and start when you want. With *Fog* you have control over the complete environment and your own motion. If your ship or space habitat were surrounded by *Fog* on the outside, it could perform tasks like maintenance, cargo stowing, and catching and docking rendezvousing craft or people. In space, matter is scarce, recycling is paramount. *Utility Fog* will be one of the most useful forms of matter, since it can do so many things. What is more, being able to change from desk to bed to instrument panel to robot to bulkhead in an instant has got to count as some kind of ultimate in recycling.

Real Virtuality

One thing in a *Fog world* that would be more difficult than ours would be telling what was real and what was not. The only technology we have that is commonly used today where this is an issue is sound reproduction—good quality recordings and speakers can be difficult to distinguish from the real source of a sound.

Utility Fog will extend this to sight and touch. So will other forms of virtual reality; but *Utility Fog* mixes virtual and real in a way that most other forms don't. What's more, with *Utility Fog* there is a philosophical conundrum to worry about as well. If there's a *Utility Fog* simulacrum of you in my living room, it clearly is not a real person. Today, we don't have any problem saying "I heard so-and-so," after speaking with him on the telephone, but we are much less likely to claim to have seen him after a video teleconference. This is not a difference between audio and video but a subtle difference in the meanings of *hear* and *see*—they are not strictly parallel. Seeing has a stronger implication of being present. We do not notice these subtleties in normal conversation but we use then nevertheless.

The bàtch of *Fog* at the end of a telepresence link looks and acts like a person, but you know there's a real person somewhere whose actions it is just copying. What if, however, the *Fog* is a *robot*, running control programs right there on the processors of the *Foglets*? Surely the hatch of *Fog* is now a "*real*" robot. Go one step further and suppose someone uploads his brain, which is then down- loaded into the *Fog robot*. Nothing else exists to be the "real" person now—its the hatch of *Fog* or nothing. It gets even worse if the down load is in *Fog* mode, not associated with any specific group of *Foglets* but a moving pattern of computation.

Our language is going to have to grow to keep up with reality. It always has. We already needed this to some extent with television and computers. A picture of a bear on television is not a real bear, but *Hamlet* read on the computer screen is really *Hamlet*. The difference does not seem to bother anybody too much, and I am sure that as we gain experience with the partly virtual world that is coming, we'll invent whatever new words and concepts are necessary to deal with it.

INDEX